Joel Dorman Steele

A Popular Chemistry

Joel Dorman Steele

A Popular Chemistry

ISBN/EAN: 9783744674959

Printed in Europe, USA, Canada, Australia, Japan

Cover: Foto ©berggeist007 / pixelio.de

More available books at **www.hansebooks.com**

J. Norman Steele.

A

POPULAR

CHEMISTRY

BY

J. DORMAN STEELE, Ph.D.

AUTHOR OF A SERIES IN THE NATURAL SCIENCES

Copyright, 1887, by

A. S. BARNES & COMPANY

NEW YORK AND CHICAGO

PUBLISHERS' PREFACE.

SINCE the publication of the revised edition of Steele's "FOURTEEN WEEKS IN CHEMISTRY," more than a decade ago, the study has grown greatly in popularity in the schools of this country. Under such circumstances, it has seemed advisable to enlarge somewhat upon the former treatment of the subject; and to meet this change of feeling on the part of teachers, the present work has been provided.

The simplicity of statement and clearness of method which are the characteristics of Professor Steele's previous text-books in Chemistry, and of all his other works, have been fully demonstrated by an ever-increasing popular demand. No change in method of treatment has been made in the present work, beyond that necessitated by new discoveries in this branch of science.

In its new form, there will be found many graphic illustrations, made expressly for this edition. The typographical appearance of the book has also been greatly improved. We trust that we have prepared a text-book that will meet the wants of both teachers and pupils.

JULY, 1887

PREFACE

TO THE FIRST EDITION.

IN the preparation of this little volume the author lays no claim to originality: his has been the far humbler task of endeavoring to express, in simple, interesting language, a few of the principles and practical applications of Chemistry. There is a large class of pupils in our schools who can pursue this branch only a single term, the time assigned to it in most institutions. They do not intend to become chemists, nor even professional students. If they wander through a large text-book, they become confused by the multiplicity of strange terms, which they can not tarry to master, and, as the result, too often only "see men as trees walking." Attempts have been made to reach this class by omitting or disguising the nomenclature; but this robs the science of its mathematical beauty and discipline, while it does not fit the student to read other chemical works or to understand their formulæ. The author has tried to meet this want by omitting that which is perfectly obvious to the eye—that which everybody knows already—that which could not be long retained in the memory—and that which is essential only to the chemist. He has not attempted to write a reference-book, lest the untrained mind of the learner should become clogged and wearied with a multitude of detail. He has sought to make a pleasant study which the pupil can master in a single term, so that all its truths may become to him " household words." Botany, Natural Philosophy, and Physiology are omitted, since they are now pursued as separate branches. Unusual importance is given to that practical part of chemical knowledge which concerns our every-day life, in the hope of bringing the school-room, the kitchen, the farm, and the shop in closer relationship. This work is designed for the instruction of youth, and for their sake clearness and simplicity have been preferred to recondite accuracy. If to some young man or woman it becomes the opening door to the grander temple of Nature beyond, the author will be abundantly repaid for all his toil.

PREFACE

SIX years ago, at the solicitation of his fellow-teachers, the author offered this work to the profession. Having been prepared for the use of his own classes, and embodying his oral instructions, it naturally partook of the peculiarities of that method. The desire was to interest pupils in scientific study. He believed that a chemical fact is no less a truth because made attractive by an imaginative garb. If thus a child could be won to its consideration, the intrinsic beauty of the subject would lure him on, and so at last he would come to pursue it into the labyrinths of dry, technical works.

The hearty reception of the book at once and its constantly increasing sale, the demand for an entire series on the same plan, words of approval from educators whose commendation it was a great satisfaction to have won, the fact that several other series based upon the same general idea have since appeared, and, above all, the assurance that the books have gone into hundreds of schools where science had never been taught before—have convinced the author of the inherent correctness of his view.

A demand having arisen for the admission of the new nomenclature into the book, the opportunity is gladly taken of making such revision as the daily use of the work in the class-room, and the advice of others, have suggested.

The author would here acknowledge his special indebtedness to the many teachers who, sympathizing with his plan of popularizing science, have pointed out what they considered defects in its execution, and given him the benefit of such illustrations and methods as they have found serviceable. The value of these criticisms has been shown in the increased worth of each edition of this series.

The usual authorities have been freely consulted in this revision. The following have been found of especial service: Miller's Elements of Chemistry (4th London Edition), Tomlinson's Miller's Inorganic Chemistry, Roscoe's Lessons in Chemistry (London, 1869), Bloxam's Metals, and Fownes' Manual of Chemistry (London, 1873). In addition, reference has been had to the works of Cooke, Draper, Nichols, Fresenius, Muspratt, Faraday, Watts, Stockhart, Moffit, Gmelin, Griffin, Tyndall, Odling, Noad, Williamson, Wilson, Galloway, Youmans, Regnault, Thomson, Valetin, Gregory, Porter, Will, and many others.

SUGGESTIONS TO TEACHERS.

IT is advised that in the use of this book the top-ical method of recitation should be adopted. So far as possible, the order of the subjects is uniform —viz., Occurrence, Preparation, Properties, Uses, and Compounds. The subject of each paragraph indicates a question which should draw from the pupil the substance of what follows. At each recitation the scholar should be prepared to explain any point passed over during the term, on the mention of its title by the teacher. Such reviews are of incalculable value. While some are reciting, let others write upon specified topics at the blackboard, after which the class may criticise the thought, the language, the spelling, and the punctuation. *Never allow a pupil to recite a lesson,* or answer a question, except it be a mere definition, *in the language of the book.* The text is designed to interest and instruct the pupil; the recitation should afford him an opportunity of expressing what he has learned, in his own style and words. Every pupil should keep a note-book, in which to record under each general head of the text-book all the experiments, descriptions, and general information given by the teacher in class. In order to accustom the scholar to the nomenclature, use the symbols constantly from the beginning; they may seem dull at first, but if every compound be thus named, a familiarity with chemical language will be induced that will be as pleasing as it will be profitable. If time will admit, in addition, have weekly essays prepared by the class, combining information from every attainable source.

Ocular demonstration is absolutely necessary to any progress in the study of chemistry. Simple directions with regard to the experiments are given in the Appendix (see page 261) which will enable the unprofessional chemist to perform them readily, and, in case it is convenient for the pupils to work in the laboratory, will guide them in their investigations. The subject of *Qualitative Analysis* is also explained so clearly, and the directions are so complete (see page 288), that even the amateur student can grasp the subject and demonstrate its principles.

Teachers desiring pleasant information to relieve the recitation hour, will find it in that delightful work of Dr. Nichols, *Fireside Science.* Many curious and entertaining stories and facts are given in a book entitled *Treasures of the Earth.* For common works of reference, Roscoe & Schorlemmer's *Treatise on Chemistry*, 6 vols. octavo, or Miller's *Elements of Chemistry*, 3 vols. octavo, will be most generally useful. These books may be obtained of the publishers of this series.

TABLE OF CONTENTS.

I.—INTRODUCTION.

II.—INORGANIC CHEMISTRY.

1.—THE NON-METALS.

III.—ORGANIC CHEMISTRY.

IV.—APPENDIX.

I.

INTRODUCTION.

" I sympathize with that beautiful idea of Oersted, which he expressed in the now familiar phrase, ' *The laws of Nature are the thoughts of God.* * * * Through the great revolutions which have taken place in the forms of thought, the elements of truth in the successive systems have been preserved, while the error has been as constantly eliminated: and so, as I believe, it always will be, until the last generalization of all brings us into the presence of that law which is indeed the thought of God."

J. P. Cooke, Jr.

AVOGADRO'S LAW.

" Equal volumes of all substances, when in a state of gas, and under like conditions, contain the same number of molecules." *See Physics, p. 23.*

ELEMENTS OF CHEMISTRY.

INTRODUCTION.

Chemistry treats of the composition of substances, and all changes in composition which may take place.

All the changes, or *phenomena*, which matter exhibits, may be divided into two groups: 1st. Those in which the composition is not altered. 2d. Those in which a change in composition takes place. The first are *physical phenomena*, the second are *chemical phenomena*.

Examples: 1st. Fall of a stone, vibration of a tuning-fork, and all phenomena of motion; boiling of water, magnetizing of iron. 2d. Rusting of iron, souring of milk, burning of candle.

No destruction of matter is possible. When a substance disappears, as in the boiling of water or the burning of a candle, this is because it is changed into an invisible form (gas), and not because it has been destroyed. In its new form it still has the properties of *weight* and *impenetrability*, which prove it to be matter.

An Element is a kind of matter which has never been separated into other substances.—*Examples:*

gold, sulphur. The number of elements now known is about seventy.* The larger number of these are rare. A few of the elements occur naturally, but most substances are composed of two or more elements, and are therefore called *compounds*.

Compounds, in. their properties, are in general very unlike their elements.†—*Examples :* yellow sulphur and white quicksilver form red vermilion; inert charcoal, hydrogen, and nitrogen produce the deadly prussic acid; solid charcoal and sulphur make a colorless liquid; poisonous and offensive chlorine combines with the brilliant metal sodium to form common salt.

Chemical Affinity, or **Chemism,** is the name given to that power which causes the chemical union of substances, and which holds the elements together in compounds. It acts only at insensible distances, and generally with great energy.‡ If chemism should suddenly cease, not only would all chemical action be impossible, but almost all substances would at once change their character; for all compounds would be resolved into their elements:—all water would disappear into two invisible gases, the solid rocks would fall to powder, and all animal and

---

* It is not probable that the list is complete, but we can not suppose that any very abundant element is yet to be found.

† "The elements have no more likeness to the compounds which they form than the separate letters of the alphabet have to the words which may be made from them."—MILLER.

‡ Nothing in the nature or appearance of an element indicates its chemical affinity, and it is only by trial that we can tell with what it will combine. This attraction is not a mere freak of nature, but is imparted to matter by God Himself.

vegetable substances would be changed into three gases and a substance like charcoal.

Heat and **Light** favor chemical action, and frequently develop an affinity where it seems to be wanting. The former, especially, tends to drive the elements of a compound without the range of old attractions and within that of new ones. **Electricity** is also a powerful agent in producing chemical action.—*Examples:* gun-cotton, when lying in the air, is apparently harmless, but a spark of fire will produce a brilliant flash, and cause it to disappear as a gas; nitrate of silver in contact with organic matter turns black, by the action of the light; an electric current led through acidulated water decomposes it into its constituent gases.

Solution aids in chemical change, as it permits the particles of substances to come within the range at which chemism can act.—*Example:* sodium carbonate* and tartaric acid mixed in a glass will not combine, but the addition of water will cause a violent effervescence.

Law of Definite Proportions.—Chemical combination always takes place between definite weights of substances.

Law of Multiple Proportions.—If two elements, A and B, combine in different proportions, the relative quantities of B which combine with any fixed quantity of A bear a simple ratio to one another.

Constitution of Bodies.—The phenomena of both Physics and Chemistry lead to the conclusion that

* Carbonate of Soda.

all bodies are made up of minute particles which
are never in actual contact with each other, and are
always in motion. In any given substance, all the
constituent particles are alike and of the same composition as the substance. These particles are called
*molecules. A molecule is the smallest particle of a
substance which can exist in the free state.* A physical change goes no further than the molecule, and
hence does not affect the nature of the substance;
but a chemical change is a change of substance, and
hence must consist in the breaking up of molecules
and the formation of new ones. This is explained
by the assumption that the molecules are compound,
each molecule being composed of still smaller particles. These smaller particles are called *atoms*, and
may be defined as *the indivisible constituents of molecules. They are the smallest particles of elements
which can take part in chemical changes.* Thus a
molecule consists of atoms held together by chemism.
In an element, the molecules are made up of atoms
of the same kind; in a compound, the molecules
consist of atoms of different kinds.

Atoms differ from each other in chemism, in
weight, and in valence.

The **Atomic Weight** of an element expresses the
weight of its atom compared with that of the atom
of hydrogen.

Molecular Weight is the sum of the weights of
the atoms in the molecule.

Valence* is the property of an atom by virtue of

* The property of valence is treated on page 111.

which it can hold a definite number of other atoms in combination.

Chemical Notation.—For the sake of brevity, chemists use a kind of short-hand. *An atom* is represented by the first letter of its English name. When that would produce confusion, the Latin initial is substituted, and in some cases a second letter added.—*Examples:* carbon ·and chlorine both commence with C; so the latter takes Cl for its symbol. Silver and silicon both begin with Si, hence the former assumes Ag, from its Latin name, Argentum. If more than one atom of an element is contained in a molecule of a compound, this is shown by writing the number below the symbol.—*Example:* H_2O indicates that in a molecule of water there are two atoms of hydrogen and one of oxygen.

Molecules are represented by grouping together the symbols of their constituent atoms. Such a group of symbols is called the *formula* of the molecule or substance.

Chemical action or "reaction" between substances is expressed by means of *chemical equations* which resemble those of algebra, and whose terms are molecular formulas.—*Example:* $NaCl + AgNO_3 = NaNO_3 + AgCl$. The sign + indicates mixture; the sign =, conversion into.

Nomenclature.—The elements which were known anciently retain their former names. Those discovered more recently are named from some peculiarity.—*Examples:* chlorine, from its green color; bromine, from its bad odor. The uniform termination

ium has been given to the lately found metals.—
Examples: potassium, sodium. A similarity of end-
ing in non-metallic elements indicates some analogy.
—*Examples:* silicon, boron ; iodine, bromine.

Compounds are named from their constituent
atoms. When a compound contains only two elements (a binary compound), the names of the two
are placed together and one (the non-metal) takes the
termination *ide.* Thus, potassium and iodine form
the compound which is written KI, and read potassium iodide ; sodium and chlorine, NaCl, sodium
chloride ; zinc and oxygen, ZnO, zinc oxide.

Other rules for nomenclature will be noticed as
occasion demands.

Classification of the Elements.—The elements are
usually divided into two classes : *Metals* and *Non-
metals.* The metals, as a class, are electro-positive *
with reference to the 'non-metals, or what amounts
to the same thing, the non-metals are electro-negative* in their behavior toward the metals. The
metals are the base†-forming elements, the non-
metals the acid†-forming elements. These classes,
however, are not separated by any sharply defined
difference in properties, but one shades gradually
into the other :—all the elements may be arranged
in a series in such a way that each is electro-positive toward all which follow it, and electro-negative
toward all which precede it ; and certain elements

* See definition of these terms under Electricity in Steele's "Popular
Physics."
† See explanation of these terms on pages 98 and 99.

are found to form both acids and bases. Thus the division is a somewhat arbitrary one, and is retained chiefly for convenience of study.

Organic and Inorganic Chemistry.—The division of Chemistry into organic and inorganic Chemistry is still kept, though the significance of the names has changed. It was formerly thought that "organic" substances could be produced only by the agency of plant or animal life, and thus formed a group quite distinct from the "inorganic" or mineral substances. But it has been found that many "organic" substances can be made in the laboratory from "inorganic" substances without the aid of the vital process. The organic substances always contain carbon, and include most of the compounds into which this element enters, so that organic Chemistry is now defined as the *Chemistry of the Compounds of Carbon*, while inorganic Chemistry deals with the compounds of the other elements.

II.

INORGANIC CHEMISTRY.

———•◆•———

"In the de-oxidation and re-oxidation of the hydrogen in a single drop of water, we have before us, so far as force is concerned, an epitome of the whole of life."—HINTON.

INORGANIC CHEMISTRY.

THE NON-METALS.

OXYGEN.

Symbol, O.....Atomic Weight, 16.....Specific Gravity, 1.1.

THE name Oxygen means acid-former, and was given because it was supposed to be the essential principle of all acids.

Occurrence.—O is the most abundant of all the elements—comprising by weight $\frac{8}{9}$ of the water, $\frac{3}{4}$ of all animal bodies, about $\frac{1}{2}$ of the crust of the earth, and more than $\frac{1}{5}$ of the air.

Preparation.—Although O is present in the air in large quantities, it can not be readily obtained from this source, but is usually prepared from one of its compounds by heat. Thus, if oxide of mercury is heated, it yields O and mercury. We may represent the chemical change which takes place by a chemical equation :-

$$HgO = Hg + O ;$$

which is read, oxide of mercury is converted into, or gives, mercury and oxygen. The equation expresses more than the mere fact that this change has taken place ; for each atomic symbol stands for a definite weight of the element which it represents, and the equation consequently means, 216 parts of oxide of mercury give 200 parts of mercury and 16 parts of oxygen.

Another substance which yields O readily when heated is potassium chlorate ($KClO_3$). The best method of preparing O for experimental purposes is to heat a mixture of equal parts of this substance and manganese dioxide (MnO_2).* The manganese dioxide remains unchanged, but in some way which is not understood, causes the potassium chlorate to give off its O at a lower temperature and more regularly than when it is heated alone. The mixed substances are heated in a flask and the gas collected over water in a pneumatic trough, as shown in the illustration.

The reaction may be represented thus:

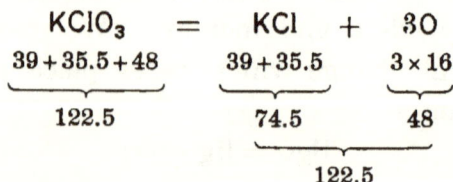

$$KClO_3 \;=\; KCl \;+\; 3O$$

$$\underbrace{39+35.5+48}_{122.5} \quad \underbrace{39+35.5}_{74.5} \quad \underbrace{3\times16}_{48}$$

$$\underbrace{}_{122.5}$$

The O obtained will be $\frac{48}{122.5}$ of the potassium chlorate used, i.e., every 122.5 parts by weight (grs., oz., or lbs.) will yield 48 parts (grs., oz., or lbs.) of O, and 74.5 parts (grs., oz., or lbs.) of KCl.

Properties.—O has no odor, color, or taste. It is but very slightly soluble in water. It combines with

---

* This substance is commonly known as binoxide of manganese, and, because of its color, the black oxide of manganese.

FIG. 1.

Collecting O over water.

every element except fluorine. From some of its
compounds it escapes explosively on the slightest
blow, while from others it can be liberated only by
the most powerful means. Its action on a substance
is called *oxidation* of the substance, and the products
are *oxides*. It is incombustible, but a vigorous supporter of combustion.

The following experiments will illustrate its chemical energy.

1. By blowing quickly upward upon a candle, extinguish the flame, and leave a glowing wick. If
this be plunged into a jar of O, the coal will burst
into a brilliant blaze. The experiment may be
repeated many times before the O will be exhausted.

A new colorless gas, CO_2, called carbon dioxide ("carbonic acid") is formed by the combustion.

FIG. 2.
Sulphur in O.

2. Ignite a bit of sulphur placed in a "deflagrating spoon" (see Appendix, p. 263), and lower it into a jar of O. It will burn with a beautiful blue light, and the formation of sulphur dioxide, SO_2 ("sulphurous acid"), which has the pungent odor of a burning sulphur match.

3. Straighten one end of a watch-spring and fasten it in a bit of thin board; heat the other end slightly and dip it into powdered

FIG. 3.
A watch-spring in O.

sulphur. Light this and plunge it into a jar of O, closing the mouth of the jar with the board. The burning sulphur will ignite the steel, which will burn without flame with a shower of fiery stars, while melted globules of the oxide of iron (Fe_3O_4) will fall upon the bottom of the jar.

4. Place in the bottom of a deflagrating spoon a little fine, dry chalk; then wipe a bit of phosphorus, about the size of a pea, very carefully and quickly between pieces of blotting-paper; lay this upon the chalk, and, holding the spoon over a large jar of O, ignite the phosphorus with a heated wire, and lower it steadily into the gas. The phosphorus will burst

into a flood of blinding light, while dense fumes of phosphorus pentoxide, P_2O_5, will roll down the sides of the jar.

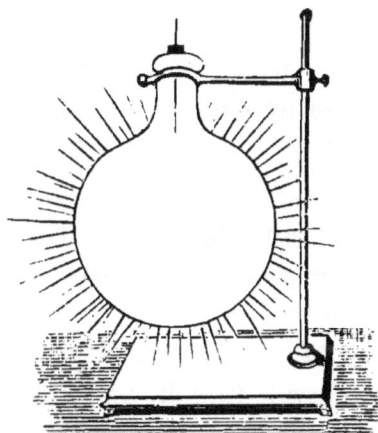

FIG. 4.

5. Make a little tassel of zinc-foil, tip the ends with sulphur as in the 3d experiment, ignite and lower into a jar of O. It will burn with a dazzling light, forming zinc oxide (ZnO).

6. If a piece of charcoal-bark be ignited and

Phosphorus in O. "The phosphoric sun."

lowered into a jar of O, it will deflagrate with bright scintillations.

Oxygen the Active Agent of the Air.—The burning of substances in air resembles the burning in O, except that it is never so energetic. It is, indeed, a true *oxidation*, and the oxides formed are the same as those produced in the O we prepare. The burning candle gives off CO_2, sulphur SO_2, phosphorus P_2O_5, the glowing iron which the blacksmith draws from his forge forms scales of Fe_3O_4, which fly blazing in every direction under the blows of his hammer.* Comprising about one fifth of the common air, O is ever-present, ever-waiting.

* Quite in contrast to this pyrotechnic display is the action of the O upon the Fe contained in writing-fluid. At first the words are pale and indistinct, but in a few hours the O, noiselessly combining with the metal (see p. 17), brings out every letter in clear, bold characters upon the page.

We open the damper of the stove and the air rushes in. The O immediately attacks the heated fuel. Every two atoms combine with an atom of C and fly off into the air as CO_2.

The water of a river becomes foul from the discharge of drains and sewers. As it flows along, exposed to the air, the O dissolves in it, attacks each particle of organic impurity and slowly burns it up; thus rendering the river-water once more fit for use.

We wipe our knives and forks, and lay them carefully away; but if we have left on them a particle of moisture, since H_2O favors chemical change, the O will find it, and corrode the steel.*

An animal dies, and the O is an important agent in removing its body. The molecules which have been used to perform the functions of life, are broken up by the O, and their atoms enter into new combinations.

O in the Human System.—We take the air into our lungs. Here the blood † absorbs the O, and bears it to all parts of the body, depositing it wherever it is needed. Laden with this life-giving element, the vital fluid sweeps tingling through every artery to the remotest capillary tubes, sends the quick flush to the cheek, combines with a portion of the

---

* The compound here formed will be a higher oxide than that produced at the blacksmith's forge, since a portion of the O was there prevented from uniting with the iron by the heat. It will be the red oxide of iron (Fe_2O_3, ferric oxide), or common iron rust, as we see it on stoves and other utensils.

† "The blood is full of red corpuscles or cells containing Fe. These are so tiny, that a million of them cluster in the drop which will cling to the point of a needle. Quickly assuming a tawny hue, like the decayed leaves of autumn, they change so rapidly that 20,000,000 perish with every breath."
—DRAPER.

food thrown into the circulation from the stomach, breaks up every worn-out tissue, burns up the muscles as they do their work, until at last it comes back through the veins dark and thick with the products of the combustion—the cinders of the flameless fire within us.

Combustion and Heat.—All ordinary processes of decay and fire, are produced by the action of O on substances, and are different forms of oxidation. They differ only in the time employed in the operation. If O unites rapidly, we call it fire; if slowly, decay. Yet the process and the products are the same. A stick of wood is burned in the stove, and another rots in the forest, but the chemical change is identical. In the oxidation of an atom of C, a certain amount of heat is produced. Hence, the house that decays in fifty years, gives out as much heat during that time as if it had been swept off by a fierce conflagration in as many minutes.

The Igniting Point of any substance is the temperature at which "it catches fire." We elevate the heat of a small portion to the point of rapid union with O, and that part in burning will give off heat enough to support the combustion of the rest.— *Example:* In making a fire, we take paper or shavings, which, being poor conductors of heat, and exposing a large surface to the action of O, are easily raised to the required temperature. Having thus obtained sufficient heat to start the combustion of chips or pine sticks, we gradually increase it until there is enough to ignite the coal or wood.

Extinguishing Fires.—Blowing on a candle or lamp extinguishes it, because it lowers the heat of the flame below the igniting point of the gases. Fires are put out by water partly for the same reason, and also because it envelops the wood and shuts off the air. If a person's clothes take fire, the best course is to wrap him in a blanket, carpet, coat, or even in his own garments. This smothers the fire by shutting out the O. Great care should be taken in a fire not to open the doors or windows, so as to cause a draught of air. The entire building may burst into a blaze, when the fire might have languished for want of O, and so have been easily extinguished.

Spontaneous Combustion.—Sometimes substances absorb oxygen from the air so rapidly, that heat enough is evolved to cause ignition; or if the substances are incombustible, other bodies in contact with them may be kindled.

The waste cotton used in mills for wiping oil from the machinery, when thrown into large heaps, often absorbs O from the air so rapidly that it bursts into a blaze. Fires are often started in this way, both in manufactories and on board ship. Similar cases of spontaneous combustion occur in hay-ricks in which the hay has been put up damp.

Heaps of coal take fire from the oxidation of the iron pyrites contained in them. This is favored by the moisture of the air.

All supposed instances of spontaneous combustion in the human body have been proved to be mistakes or deceptions.

The Human Furnace.—The body is like a stove in which fuel is burned, and the chemical action resembles that in any other stove. This combustion produces heat, and our bodies are kept warm by the constant fire within us. We thus see why we fortify ourselves against a cold day by a full meal. When there is plenty of fuel in our human furnaces, the O burns that; but if there is a deficiency, the destructive O must still unite with something, and so it combines with the flesh :—first the fat, and the man grows poor; then the muscles, and he grows weak; finally the brain, and he becomes crazed. He has burned up, as a candle burns out to darkness.

O Produces Motion.—As soon as we begin to perform any unusual exercise, we commence breathing more rapidly, showing that, in order to do the work, we need more O to unite with the food * and muscles. In very violent labor, as in running, we are compelled to open our mouths, and take deep inspirations of O. This increased fire within elevates the temperature of the body, and we say "we are so warm that we pant." Really it is the reverse. The panting is the cause of our warmth.

During sleep the organs of the body are mostly at rest, except the heart. To produce this small muscular exertion very little O is required. As our respiration is, therefore, slight, our pulse sinks, the

---

* It is probable that a portion of our food, especially the carbonaceous, is oxidized directly without becoming an integral part of the body. The heat thus set free may, by the principle of the *Conservation of Energy*, be converted into muscular energy.

heat of our body falls, and we need much additional clothing to keep warm.* Thus we require O not only to keep us warm, but also to do all our work. Cut off its supply, and we grow cold; the heart struggles spasmodically for an instant, but the motive power is gone, and we soon die.

How O Gives us Strength.—Our muscles, as well as the food from which they are formed, consist of complex organic bodies, and the pent-up energy is very great. Thus in flesh, starch, sugar, etc., the molecules are very complex (see p. 188), and when these oxidize into the simpler ones of water, carbonic acid, and ammonia, the potential energy is transformed into heat and muscular strength.† As no matter is either lost or gained in any chemical change, so also no energy is lost or gained, but all must be accounted for.

The Burning of the Body by O.—A man weighing 150 lbs. has 64 lbs. of muscle. This will be burned in about 80 days of ordinary labor. As the heart works day and night, it burns out in about a month. So that we have a literal "new heart" every thirty days. We thus dissolve, melt away in time, and only the shadow of our bodies can be called our

---

* Animals that hibernate show the same truth. The marmot, for instance, in summer is warm-blooded; in the winter its pulse sinks from 140 to 4, and its heat corresponds. The bear goes to his cave in the fall, fat; in the spring he comes out lean and lank. Cold-blooded animals have very inferior breathing apparatus. A frog, for example, has to swallow air by mouthfuls, as we do water. Others have no lungs at all, and breathe in a little air through the skin, enough barely to exist. Is it strange they are cold-blooded?

† See discussion of energy and the transformation of potential into kinetic energy in "Steele's Popular Physics."

own. They are like the flame of a lamp, which appears for a long time the same, since it is "ceaselessly fed as it ceaselessly melts away." The rapidity of this change in our bodies is remarkable. Says Dr. Draper : "Let a man abstain from water and food for an hour, and the balance will prove he has become lighter." This action of O, so destructive— wasting us away constantly from birth to death— is yet essential to our existence. Why is this? Here is the glorious paradox of life. *We live only as we die.* The moment we cease dying, we cease living. All our life is produced by the destruction of our bodies. No act can be performed except by the wearing away of a muscle. No thought can be evolved except at the expense of the brain. Hence the necessity for food to supply the constant waste of the system,* and for sleep to give nature time to repair the losses of the day. Thus, also, we see why we feel exhausted at night and refreshed in the morning.

O the Common Scavenger.—God has no idlers in His world. Each atom has its use. There is not an extra particle in the universe. The mission of oxygen, so destructive in its action, is therefore essential, that every waste substance may be collected and returned to the common stock, for use in nature's laboratory. In performing this general task, its uses

---

* This food must be organic matter endowed with potential energy treasured up in the plant. When it is transformed into flesh, perhaps made still more vital in the process, we have this power standing ready to be used again at our pleasure. When we will it, the O combines with the flesh and sets free the energy for us to apply.

are most important and necessary. It sweetens water, it keeps the avenues of the body open and unclogged,* it preserves the air wholesome. It becomes, in a word, the universal scavenger of nature. Every dark cellar of the city, every recess of the body, every nook and cranny of creation, finds it waiting; and the instant an atom is exposed, the oxygen seizes upon it. A leaf falls, and its destruction forthwith commences. A tiny twig, far out at the end of a limb, dies, and the O immediately begins its removal. A pile of decaying vegetables, a heap of rubbish, the dead body of an animal, a fallen tree, the houses we build for our shelter, even the monuments erected above our final resting-place, are all gnawed upon by what we call the "insatiate tooth of time." It is only the constant corrosion of this destructive agent—oxygen.

Consumption of O.—Each adult uses daily 1¼ lbs. of O. The combustion of 1 lb. of coal requires 2⅔ lbs. of O: so that the ship which burns 1,000 tons in crossing the ocean, takes out of the air 2,666 tons of O. Supposing the population of the earth to be 1,200,000,000, and each* person to consume 1 lb. of O; adding as much more to sustain fires; twice as much for the wants of animals, and four times as much for the varied processes of decay, the daily consumption of O reaches the enormous sum of 4,800,000 tons (Faraday). Yet the atmosphere contains over one quadrillion tons, and

---

* Huxley very prettily calls O, in this connection, the " great sweeper " of the body, since it lays hold of all the waste matter of the system, and burning it up, removes it out of the way.

even this vast aggregate is a mere fraction compared with the O locked up in the ocean and the rock.

Results if the Air were Undiluted O.—The fire element would run riot· every-where. Metal lamps would burn with the oil they contain. Our stoves would blaze with a shower of sparks. A fire once kindled would spread with ungovernable velocity, and a universal conflagration would quickly wrap the world in flame.

OZONE.

Ozone is an *allotropic* form of O—*i.e.*, a form in which the element itself is so changed as to have new properties.

Source.—It is always perceived during the working of an electric machine, and is then called "the electric smell." It is also said to have been noticed during thunder-showers, and is formed by evaporation and various processes of combustion.

Fig. 5.

Testing ozone.

Preparation.—Place a freshly scraped stick of phosphorus in a jar containing a little water, so that it shall be partly covered with the water. It will slowly oxidize, and the peculiar odor of ozone will soon be perceived in the jar. It may also be tested by a paper wet with a mixture of starch and potassium iodide (KI). The ozone sets free the iodine, which unites with the starch, forming blue iodide of starch.* At a temperature above that of boiling water, the ozone will turn back into O.

Properties.—Ozone is still more corrosive than oxygen. It tarnishes mercury, which oxygen does not attack at ordinary temperatures; it bleaches powerfully, and it is a rapid disinfectant. A piece of tainted meat plunged into a jar of it is instantly deodorized, and it is probable that, even in minute quantities, this gas exercises a powerful influence in purifying the atmosphere. Ozone is condensed oxygen. Its molecule consists of three atoms of O instead of two, as in ordinary oxygen. The change of oxygen into ozone may be thus represented:

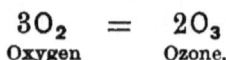

$$3O_2 \quad = \quad 2O_3$$
$$\text{Oxygen} \qquad \text{Ozone.}$$

WEIGHING AND MEASURING GASES.

It is so much easier to measure the volume of a gas than it is to weigh it, that in practice the gas obtained by any reaction is usually measured and its amount or weight calculated from its volume. But since a given quantity of any gas by weight occupies a different volume at different temperatures and at different pressures, both temperature and pressure (height of barometer) at the time of measuring, have to be taken into account in making the calculation. The law for the change in volume produced by tem-

* If a piece of the dry iodized paper be exposed upon a clear day to the open air of the country, in a few minutes it will assume a bluish tint. In cloudy, foggy weather, or in cities, this effect is rarely observed.

perature is :—*a gas expands or contracts $\frac{1}{273}$ part of the volume it would have at* 0° C., *for every change of one degree centigrade in its temperature.* Hence, if V be the volume of a gas measured at 0° C., and v the volume it would assume at $t°$ C., $v = V + \dfrac{t}{273} \cdot V$. If v be the volume at $t°$ C., the volume at 0° C., or

$$V = v\frac{273}{273 + t}.$$

The law for the effect of pressure is :—*the volume of a gas varies inversely as the pressure.* Hence, if the volume is V under pressure P and v under pressure p, $V : v :: p : P$; and $V = \dfrac{vp}{P}$ or $v = \dfrac{VP}{p}$.

The pressure is stated in millimeters, and refers to the height of a column of mercury (barometer) which the pressure would sustain. 760 *mm.* is taken as the standard pressure, and 0° C. as the standard temperature.

When the weight of a liter of any gas under standard conditions is known, the weight of any volume measured under any known conditions may be readily calculated. The steps of the calculation are: 1st. Find what the volume would be under standard conditions by means of the formulas given above. 2d. Multiply this result by the weight of one liter of the gas under these standard conditions.

By means of our chemical equations we can find the weight of a gas which a given weight of the re-agents will produce (see p. 12). Now, the volume which this weight of gas will occupy under any pressure and at any temperature, may be found by the following steps:—1st. Divide the weight of the gas by the weight of our "standard" liter—this gives the volume under standard conditions. 2d. Find the volume at the given temperature and pressure by the use of the formulas.

In this way we can find out exactly how many liters of a gas (*e.g.*, O), measured at the temperature of the room and under the existing atmospheric pressure, can be obtained from a given weight of the substance used for its production (*e.g.*, $KClO_3$).

PRACTICAL QUESTIONS.

1. What becomes of the water that "dries up"? Of the wood that "burns up"? Is there any destruction of the matter they contain?
2. Where is the higher oxide formed, at the forge or in the pantry?
3. Why is the blood red in the arteries, and dark in the veins?
4. Do we need more O in winter than in summer?
5. Which would starve sooner, a fat man or a lean one?
6. How do teamsters warm themselves by slapping their hands together?
7. Could a person commit suicide by holding his breath?
8. Why do we die when our breath is stopped?
9. Why do we breathe so slowly when we sleep?
10. How does a cold-blooded animal differ from a warm-blooded one?
11. Why does not the body burn out like a candle?
12. Do all parts of the body change alike?
13. What objects would escape combustion if the air were undiluted O?
14. Why is it difficult to obtain O from the air?
15. What weight of O can be obtained from 10 grams of HgO?
16. How much O can be obtained from 6 grams of $KClO_3$?
17. How much $KClO_3$ would be needed to produce 2 kilograms of O?
18. How much KCl would be formed in preparing 1 kilogram of O?
19. Is it probable that all the elements are discovered?
20. Is heat *produced* by oxidation?
21. What is the difference between kinetic and potential energy?
22. Why does running cause panting?
23. How does O give us strength?
24. Does the plant *produce* energy?
25. If we burn an organic body in a stove it gives off heat; in the body it produces also motion. Explain.

26. Why does not blowing *cold* air on a fire with a bellows extinguish it?

27. Why does blowing on a fire kindle it, and on a lighted lamp extinguish it?

28. Why can we not ignite hard coal with a match?

29. Why will an excess of coal put out a fire?

30. Could a light be frozen out, *i. e.*, extinguished, by merely lowering the temperature?

31. Why is it beneficial to stir a wood-fire, but not one of anthracite coal?

32. Why will water put out a fire?

33. What should we do if a person's clothes take fire?

34. Ought the doors of a burning house to be thrown open?

35. How much O can be obtained from 100 grams of HgO?

36. What would be the volume of the O of Question 35 under the standard conditions?*

37. What would be the volume of the O at 12° C, and under a pressure of 740 *mm.* of mercury?

38. What would be the volume of the O of Question 16 at 20°C. and 750 *mm.*?

39. How much KClO₃ must be employed to make an amount of O which shall measure 100 liters at 18° C. and 760 *mm.*?

---•••---

NITROGEN.

Symbol, N . . . . . Atomic Weight, 14 . . . . . Specific Gravity, 0.97.

THIS gas is called nitrogen because it exists in niter.

Occurrence.—N forms about ⅘ of the atmosphere, and is found abundantly in nitrates (*e. g.*, saltpeter), ammonia, flesh,† and in such vegetables as the mushroom, cabbage, horse-radish, etc. It is an essential constituent of the valuable medicines, quinine and morphine, and of the potent poisons, prussic acid and strychnine.

* The weight of a liter of O under standard pressure and at standard temperature is 1.43 gram.

† Its compounds give to burnt hair and woolen their peculiar odor.

Preparation.—As the air consists almost exclusively of N and O, the easiest method of obtaining the former gas is to remove the latter by employing it to oxidize some substance. The substance should be one which forms an oxide which is solid, liquid, or readily soluble,

Fig. 6.

Preparing N.

so that no gaseous product may be left mixed with the N. Place in the center of a deep dish of water a little stand several inches in height, on which a bit of phosphorus may be laid and ignited. As the fumes of phosphorus pentoxide ascend, invert a receiver over the stand. The phosphorus will consume the O of the air contained in the jar, leaving the N. After the jar has cooled it will be found that the N occupies $\frac{4}{5}$ of the receiver. The jar will at first be filled with white fumes (P_2O_5), but they will be absorbed by the H_2O in a short time.

Properties.—N is of an entirely negative character. It is colorless, odorless, and tasteless. It neither burns nor permits any thing else to burn. A candle will not burn in it, and a person can not breathe it alone and live, simply because it shuts off the life-giving O. So will a person drown in H_2O, not that the water poisons him, but because it fills his mouth, and shuts out the air N has only a weak affinity for any of the elements. The instability of its compounds is a striking peculiarity. It will unite with iodine, for example, but a brush with a feather, or a heavy step on the floor, will set it free.*

Uses.— Relation of N to Organic Substances. — Four fifths of each breath that enters our lungs is N ; yet it comes out as it went in,† while that portion of the O which remains behind performs its wonderful work within our bodies. About one sixth of our flesh is N, yet none of it comes from the air we breathe. We obtain all our supply from the lean meat and vegetables we eat. Plants breathe the air through the leaves—their lungs ; yet they do not appropriate any of the N obtained in this way, but rely upon the ammonia and the nitric acid their roots absorb from the soil. N enters the stove with the O—the latter unites with the fuel ; but the former, having no chemical attraction, passes out of the chimney. Even from a blast-furnace, where Fe melts like wax, N comes forth without the smell of fire upon it (p. 152). So inert is it, that it will not unite directly with any organic substance. We must all, animals and plants, depend upon finding it already combined in some chemical compound, and so appropriate it to our use. But even then we hold it very

---

* "Like a half-reclaimed gypsy from the wilds, it is ever seeking to be free again; and not content with its own freedom, is ever tempting others, not of gypsy blood, to escape from thralldom. Like a bird of strong beak and broad wing, whose proper place is the sky, it opens the door of its aviary, and rouses and flutters the other and more peaceful birds, till they fly with it, although they soon part company."—*Edinburgh Review*.

† There is a constant though minute exhalation of N through the pores of the skin. This small amount is perhaps absorbed in the lungs, but it is of no use to the body, so far as known.

loosely indeed. The tendency of flesh to decompose is largely owing to the instability of nitrogen compounds.

Difference between N and O.*—We see, now, how different N is from O. The one is the conservative element, the other the radical. But notice the nice planning shown in the adaptation of the two to our wants. O, alone, is too active, and must be restrained; N, alone, is sluggish, and fit only to weaken a stronger element. Were the air undiluted O, our life would be excited to a pitch of which we can scarcely dream, and would sweep through its feverish, burning course in a few days; were it undiluted N we could not exist a moment. Thus we see that, separately, either element of the air would kill us, O by excess and N by lack of action.

O and N combined.—A mixture of the active O and the inert N gives us the golden mean. The O now quietly burns the fuel in our stoves and keeps us warm; combines with the oil in our lamps and gives us light; acts upon the materials of our bodies and gives us warmth and strength; cleanses the air and keeps it fresh and invigorating; sweetens foul water and makes it wholesome; works all around and within us a constant miracle, yet with such delicacy and quietness that we never perceive or think of it until we see it with the eye of science.

Compounds. — Nitric Acid, HNO_3. — Sources. —This

---

* The difference between these two gases can be best illustrated by having a jar of each, and rapidly passing a lighted candle from one to the other; the N will extinguish the flame, and the O relight the coal. By dexterous management, this may be repeated a score of times.

compound of H, N, and O can not be easily made by the direct union of its elements, on account of the inert character of N; but compounds closely allied to it are formed in favorable soils by the decomposition of the waste products of animal life. These compounds contain a metal, usually K or Na, in place of the H, and are called *nitrates*—*e. g.*, KNO_3, potassium nitrate; $NaNO_3$, sodium nitrate.

FIG. 7.

Preparing HNO_3.

Preparation.—Nitric acid is prepared by treating a nitrate with a stronger acid. Thus, if sulphuric acid (H_2SO_4) and sodium nitrate be gently heated together in a retort, nitric acid will distil over and can be collected in a receiver, cooled by dripping water.

The chemical reaction may be represented thus:

$$2NaNO_3 + H_2SO_4 = Na_2SO_4 + 2HNO_3.$$

Properties.—It is an intensely corrosive, poisonous liquid. When pure, it is colorless; but as sold, it has commonly a golden tint from the presence of an oxide of N, produced by the decomposing action of the light. In strength it is next to H_2SO_4. It dissolves most metals with the formation of nitrates. It was formerly called *aqua fortis*, or strong water. It stains wood, the skin, etc., a bright yellow. Both nitric acid and nitrates give up their O readily, and hence are powerful oxidizing agents.*

Uses.—HNO_3 is employed in dying silk yellow, in making gun-cotton, nitro-glycerin, and other explosives, and in surgery for cauterizing the flesh. In combination with HCl, it forms *aqua regia*, the usual solvent of Au. It etches the lines in copper-plate engraving, and the beautiful designs on the blades of razors, swords, etc. The process is very simple: the surface is covered with a varnish impervious to the acid, and the desired figure is then sketched in the varnish with a needle. The HNO_3 being poured on, dissolves the metal in the delicate lines thus laid bare.

Nitrous Oxide, N_2O.—Preparation.—This gas is made by heating ammonium nitrate (NH_4NO_3), which decomposes into $2H_2O$ and N_2O. (See p. 33.)

Properties.—N_2O is a colorless, transparent gas

---

* The following experiments illustrate this property:

1. Mix equal parts of strong HNO_3 and H_2SO_4. Place a little oil of turpentine in a cup out-of-doors, and pour the mixture upon it at arm's length. The turpentine will burn with almost explosive violence.

2. Pour dilute HNO_3 upon bits of tin. Dense, red fumes (NO_2, nitric peroxide) will pass off, and the Sn will be converted into a white oxide, which furnishes what is termed putty powder.

3. Throw crystals of any nitrate on red-hot coals. They will deflagrate on account of the O which they give up to the fire.

4. Soak a strip of blotting-paper in a solution of niter. It will form "touch-paper," and when lighted will only smolder.

with a faintly sweetish taste and smell. It is somewhat soluble in water, so that there is some loss
when it is collected over water. It supports combustion nearly as
well as O, and many
of the experiments
ordinarily per-
formed with that
gas will be equally
brilliant with N_2O.
If breathed for a
short time, it pro-
duces a peculiar
kind of intoxica-
tion, often attended

FIG. 8.

Preparing N_2O.

with uncontrollable laughter, and hence it has received the popular name of *laughing gas*. The effect
soon passes off. If taken for a longer time, it causes
insensibility, and is therefore valuable as an anæsthetic in minor surgical operations, as in pulling teeth.

Nitric Oxide, NO.—Preparation.—This gas may be
prepared by the action of dilute HNO_3 on copper
clippings. The flask (*a*, Fig. 9) will soon be filled
with red fumes, but a colorless gas will collect in
the jar over water. At the conclusion of the process,
the flask will contain a deep blue solution of copper
nitrate ($Cu2NO_3$). By filtering and evaporating, the
beautiful crystals of this salt may be obtained.

There are two changes involved in the reaction ;
in the first, copper nitrate is formed and H set free :

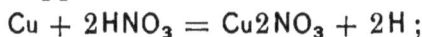

$$Cu + 2HNO_3 = Cu2NO_3 + 2H ;$$

and then the H is oxidized by the nitric acid with the production of water and NO:

$$2HNO_3 + 6H = 4H_2O + 2NO.$$

Fig. 9.

Preparing NO.

Properties.—NO is a colorless, irrespirable gas with a disagreeable odor. It does not burn, nor does it support combustion, although it contains twice as much O as N_2O. This shows that the O is held more firmly than in the latter gas. Its remarkable property is its affinity for O. Let a bubble escape into the air, and red fumes of nitric peroxide (NO_2) will be formed.*

* This may be illustrated still more prettily by the following experiment:—Fill a small jar with water colored blue by litmus solution, and pass up into it sufficient NO to occupy about one third of the bottle; the litmus will not change in color. Now allow a few bubbles of O to rise into the NO; deep red fumes will be formed, which will quickly dissolve, and the blue solution become red. If both the O and the NO be pure, it is possible, by cautiously adding O, to cause a complete absorption of both gases. If common air were used instead of O, only N would then remain in the jar.

Ammonia, NH₃.—Source.—This gas was formerly called hartshorn, because in England it was made from the horns of the hart. It received the name *ammonia*, by which it is now more generally known, from the temple of Jupiter Ammon, near which sal-ammoniac, one of its compounds, was once manufactured. The *aqua ammonia* of the shops, which is merely a strong solution of the gas in H_2O, is obtained from the incidental products of the gas-works in large quantities. (See p. 72.) Its pungent odor can often be detected near decaying vegetable and animal matter.

FIG. 10.

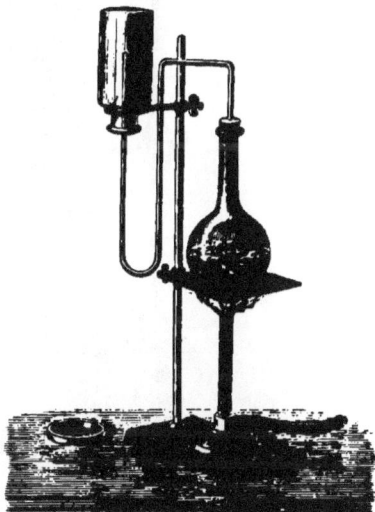

Preparing NH₃.

Preparation.—NH_3 is ordinarily prepared by heating sal-ammoniac with lime.* The reaction may be represented as follows:

$$2NH_4Cl + CaO = 2NH_3 + H_2O + CaCl_2.$$

It is also conveniently obtained for experiments by gently heating aqua ammonia.

Properties.—NH_3 is a colorless gas, having a peculiar pungent and suffocating odor, and a caustic taste.

* This may be illustrated by simply mixing in a cup some powdered sal-ammoniac (ammonium chloride) and lime (calcium oxide), when the ammonia may be detected by its odor, and the bluing of moist red litmus-paper.

It can be readily condensed to a liquid by cold or pressure. When liquefied by pressure, it passes rapidly back into the gaseous state when the pressure is removed, and in doing so, absorbs so much heat

FIG. 11.

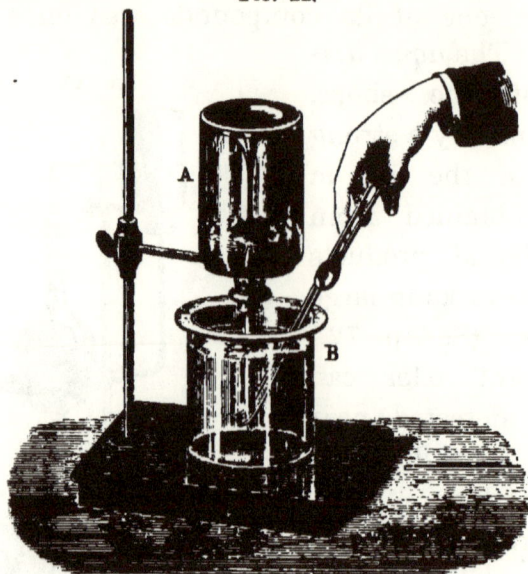

Absorption of NH_3 *in water.*

that water can be frozen.* NH_3 will not support combustion, nor will it burn in air under ordinary conditions; but it burns in O with a pale yellow flame.   It dissolves very freely in water,† forming a

* Carré's machine for making artificial ice makes use of these facts. For a description of this see Roscoe and Schorlemmer, under Ammonia.

† Heat a little aqua ammonia in a flask. Dry the vapor and collect in an inverted bottle, for which a cork and tube, with the inner extremity drawn to a fine point over the spirit lamp, has been provided. Insert the cork, and then plunge the bottle into a vessel of water. The water which passes in first will absorb the gas so quickly as to make a partial vacuum, into which the water will rush so violently as to produce a miniature fountain. If the water is colored with a little red litmus, it will turn blue as it enters the bottle.

solution which smells strongly of
the gas, and from which it can be
all driven off by heat.

FIG. 12.

Nascent State.—If N and H, the
elements of NH_3, be mixed in a
receiver, they will not unite chemically, owing to the negative character of N. When, however, any
substance is decomposed which contains both of them, as bituminous
coal, flesh, etc., at the very instant
of their separation they will combine and form NH_3. When ele-

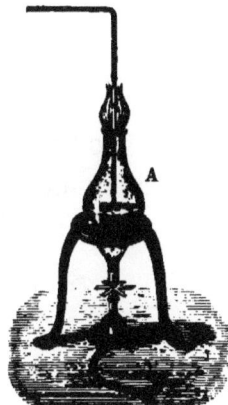

NH_3 *burning in* O.

ments are thus in the act of leaving their compounds, they are said to be in the "nascent state."

PRACTICAL QUESTIONS.

1. How could you detect any free O in a jar of N?
2. How would you remove the product of the test?
3. In the experiment shown in Fig. 9, why is the gas red in the flask, but colorless when it bubbles up into the jar?
4. How much NH_3 can be obtained from 3 grams of sal-ammoniac?
5. What will be the volume of the NH_3 at 20° C. and 770 $mm.$?
6. How much H_2O will be formed in the process?
7. How much CaO will be needed?
8. How much N_2O can be made from 1 gram of ammonium nitrate?
9. How much nitric acid can be formed from 50 kilos of sodium nitrate ($NaNO_3$)?
10. What causes flesh to decompose so much more easily than wood?
11. If a tuft of hair be heated in a test tube, the liquid formed will turn red litmus paper blue. Explain.
12. Why should care be used in opening a bottle of strong NH_3 in a warm room?
13. What weight of N is there in 10 grams of HNO_3?
14. How much sal-ammoniac would be required to make 20 liters of NH_3 measured at 25° C. and 744 $mm.$?
15. What is the difference between liquid ammonia and liquor ammoniæ?

HYDROGEN.

Symbol, H . . . . . Atomic Weight, 1 . . . . . Specific Gravity, .069.

HYDROGEN means literally a generator of water.

Occurrence. — H forms one ninth the weight of water, and is a constituent of all animal and vegetable substances.

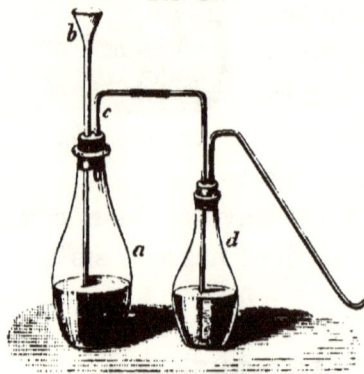

FIG. 13.

Preparing hydrogen.

Preparation. — It may be obtained from water by means of the electric current, or by the action of certain metals. If an electric current be led through acidulated water, H is given off at the negative pole and O at the positive pole. If a small piece of sodium is thrown on water, it melts and rolls over its surface like a tiny silver ball. If the water be heated, the ball bursts into a bright yellow blaze. If potassium be used instead of sodium, the H catches fire at once, even on cold water, and burns with some volatilized K, which tinges the flame with a beautiful purple tint.* If the water be examined after the action is over, it is found to feel soapy, to turn red litmus paper blue,

FIG. 14.

K *on* H_2O.

---

* Cut the metal in small pieces and cover it with a receiver, since the melted globule bursts at the close of the experiment.

and to leave, on evaporation, a white substance. This white substance is KOH or NaOH, potassium or sodium *hydroxide*. The formation of this substance and of H by the action of Na on water is represented as follows:

$$H_2O + Na = NaOH + H.$$

The H made in this way may be collected in an inverted test tube full of water by imprisoning the globule of Na in a cage of wire gauze beneath the mouth of the tube.

H is usually prepared, however, from sulphuric acid (H_2SO_4) by the action of zinc. The reaction is as follows:

$$H_2SO_4 + Zn = ZnSO_4 + 2H.$$

The $ZnSO_4$ (zinc sulphate) is contained in the solution which remains, and may be obtained in crystals by evaporating off the water.

Properties.—H prepared in this manner has a disagreeable odor, from impurities which it contains.* When pure, it is, like O, colorless, transparent, and odorless. It is the lightest of all bodies, being fourteen and a half times lighter than air, and sixteen times lighter than O. It is not poisonous, although, like N, it will destroy life by shutting out the life-sustainer, O. When inhaled, it gives the voice a ludicrously shrill tone. It can be breathed for a few moments with impunity, if it be first purified. Owing to its lightness, it passes out of the lungs again

---

* This odor can be removed by causing it to bubble through a solution of potassium permanganate. (See Fig. 13.)

directly. Its levity suggested its use for filling balloons,* and it has been employed for that purpose;
but coal gas, which contains much H and is cheaper, is now preferred.

Fig. 15.

Candle in H.

Combustion of H.—A lighted candle, plunged into an inverted jar of H, is extinguished, while the gas itself takes fire and burns with an almost invisible flame. One atom of the O of the air unites with two atoms of the H, and the product of the combustion is H_2O, which may be condensed on a cold tumbler, held over a jet of the burning gas. (See Fig. 17.) The apparatus shown in Fig. 16 is a more simple means of illustrating the properties of H.†

Fig. 16.

Mixed Gases. — A mixture of two parts, by measure, of H, with one part of O, or five parts of common air, when ignited, will explode violently.‡ The heat generated by the union of H

---

* We read in accounts of fêtes at Paris, of balloons ingeniously made to represent various animals, so that aërial hunts are devised. The animals, however, persistently insist upon ascending with their feet up—a circumstance productive of great mirth in the crowd of spectators.

† Let the gas escape a few moments before lighting it, so that the air may be driven out of the flask.

‡ The H gun—which is simply a tin tube, closed at one end, and provided with a cork at the other, having a priming-hole at the side—is used to illustrate this fact. It may be filled over the jet of the evolution flask (Fig. 16) when that is not ignited. The gas is allowed to pass in until the gun is about a fifth full, as nearly as one can guess, when the gun is removed and the gases ignited at the priming-hole.

FIG. 17.

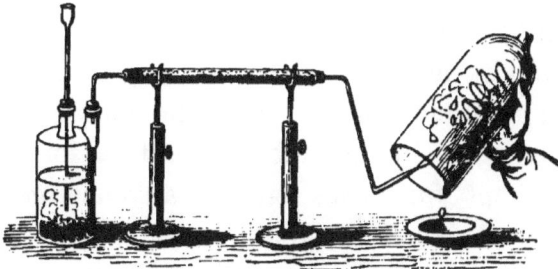

H_2O formed by burning H.

and O causes the H_2O which is formed to appear in
the state of steam. Immediately after, the steam

FIG. 18.

Transferring gases.

being condensed, a vacuum is produced and the particles of air rushing in to fill the empty space, by

their collision against each other, cause the deafening sound. While the detonation is so great, the force is slight, as may be shown by exploding, in the hand, soap-bubbles blown with the gases. H and O may be mingled in the right proportion for combustion, and kept for years without any change taking place. The two gases remain quietly together, with no manifestation of their chemical affinity, until suddenly, at the contact of the merest spark of fire, they rush together with a crash like thunder, and uniting, form the bland, passive liquid—water.

Action of Spongy Platinum.—A piece of spongy platinum placed in a jet of H will ignite it. This

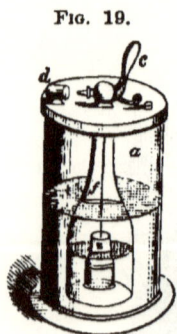

FIG. 19.

Döbereiner's Lamp.

curious effect seems to be produced in the following way: The H and the O of the air are brought so closely together in its minute pores that they unite, and the heat thus generated sets fire to the gas. This action is nicely shown by the instrument represented in Fig. 19. It was formerly used by chemists as a convenient way of obtaining a light in the laboratory. Friction matches have superseded this ingenious invention.

Heat of Burning H.—A hydrogen flame gives little light, but great heat. In H and O, existing as gases, there is stored a vast amount of potential energy.

* Z is a piece of zinc suspended in a solution of dilute H_2SO_4. At the top is a stop-cock, by turning which the gas is allowed to pass out from the receiver *f*. It strikes upon a piece of spongy platinum, and ignites with a slight explosion.

("Physics," pp. 36, 37.) When they unite by chemical affinity, this energy is transformed chiefly into heat. In the union of 8 grams of O and one gram of H, sufficient heat is evolved to raise 34,462 grams of water from 0° to 1° centigrade · and this heat is sufficient to do the amount of work represented by lifting 14,612 kilograms a meter high.

Fɪɢ. 20.

Hydrogen tones.

The Chemical Harmonica.* — The vibration of a column of air can be illustrated by simply holding a long glass tube, by means of a suitable clamp,

* Another illustration of singing hydrogen may be represented in the following manner: Make a jar of heavy tin, in the form of a double cone, twelve inches long añd four inches in diameter. At one apex fit a nozzle and cork; at the other, make several minute openings. Cover the holes with sealing-wax, and draw the cork; then fill the jar with H, and replace the cork. When ready for use, hold the jar in a vertical position, remove the wax from at least one orifice, ignite the H at that point, and draw the cork. Still hold the jar quietly, and in a minute or two the tiny jet of H will begin to sing like a swarm of mosquitoes, buzzing and humming in a most aggravating way until, unexpectedly, the fitful music ends in a loud explosion.

over a minute jet of burning H. At first no effect
will be produced; but as we slowly introduce the
jet farther and farther into the tube, a faint sound
is heard, apparently in the far-off distance. It gradually strengthens, and finally bursts into a shrill,
continuous, musical note—the key-note of the heated
column of air within the tube. The flame rises and
falls in rapid succession without ever becoming
quite extinguished, as may be seen by looking at
its image in a revolving mirror; and the succession of minute collapses of the body of air around
it is regulated by the 'length of the pipe. Let us
now place the tube at a point' where no clapping
of hands or unusual sound will start it into song.
Let various tones be produced from a violin, and
we shall find the flame responding only to that
tone which is the key-note of the tube, or its octave.
The violin player will have perfect control of this
musical flame, and can start, stop, or throw it into
violent convulsions, even across a large hall. Tubes
of different sizes and lengths will give tones of
diverse character and pitch.* The waves of sound
from the instrument augmenting or interfering
with those in the tube probably produce these phenomena.

---

* The singing of the hydrogen flame may be illustrated by holding
large tubes of any kind, over the flame of the evolution flask. Jets of different sizes may be made by drawing out glass tubing over the spirit-lamp.
Attaching a jet, by india-rubber tubing, to the nozzle of a common gas
pipe, we may utilize the H in coal gas, and at the same time secure a
brighter flame and regulate the pressure at will. The singing may be produced even if the jet and tube be horizontal or inverted.

WATER.   45

WATER.

THE COMPOSITION of H_2O is proved by analysis and synthesis—*i. e.*, by separating the compound into its elements, and by combining the elements to produce the compound. We can analyze it in the manner already shown in preparing H by passing through it an electric current. In the syn-
thetic method, we mix the two gases and unite them as we have before or by an electric spark. Both methods agree in proving that water is composed of two volumes of H to one of O. But O is sixteen times heavier than H, volume for vol-ume, and hence the composition of water by weight is 2 : 16 or 1 : 8. This fact is expressed in its formula H_2O. The black-smith decomposes water when he sprinkles it on the hot coals

FIG. 21.

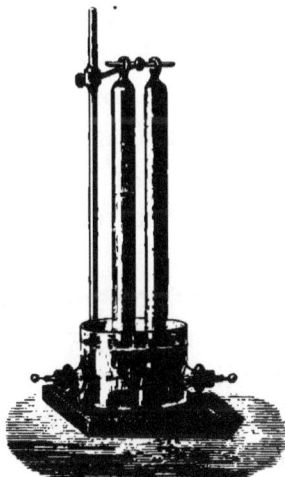

Analysis of water.

in his forge. The H burns with a pale flame, while the O increases the combustion. Thus, in a fire, if the engines throw on too little water, it may be decom-posed, and add to the fury of the flame.* To "set

* "No more heat is produced by the action of the H_2O, but it is in a more available form for communicating heat. The steam in contact with incandescent charcoal is decomposed—the O going to the C to form CO_2, and the H being set free. If the C is abundant, and the heat high, the CO_2 is also decomposed, and double its volume of CO formed. The inflam-mable gases, H and CO, mingled with the hydrocarbons always produced, are ignited, making the billows of flame which sweep over a burning building."—S. P. SHARPLES.

the North River on fire" is only a poetical exaggeration.

The quantity of electricity required to decompose a single grain of water is estimated to be equal to that in a flash of lightning. The enormous power necessary to tear these two elements from each other shows the wonderful strength of chemical attraction.* We thus see, that in a tiny drop of dew there slumbers the latent power of a thunder-bolt.

Water in the Animal World.—The abundance of water very forcibly attracts the attention. It composes perhaps four fifths of our flesh and blood. Man has been facetiously described as twelve pounds of solid matter wet up in six pails of water. All plumpness of flesh, and fairness of the cheek, are given by the juices of the system. A few ounces of water and a little charcoal constitute the principal chemical difference between the round, rosy face of sixteen, and the wrinkled, withered features of three-score and ten. To supply the constant demand of the system for water, each adult, in active exercise, needs about three pints per day, or over half a ton annually. (See "Physiology," p. 220.) When we pass to lower orders of animals, we find this liquid still more abundant. Sunfishes are little more than organized water. Professor Agassiz analyzed one found off the coast of Massachusetts, which weighed thirty pounds, and obtained only half an ounce of dried flesh. Indeed, an entire class of animals (hydrozoa),

---

* The power needed to separate them becomes latent in the gases as a potential energy, and when they are burned at any time will be set free as sensible heat—a form of kinetic energy.

to which belong the jelly-fish, medusa, etc., is composed of only ten parts in a thousand of solid matter. (See "Zoology," p. 269.)

Water in the Vegetable World.—In the vegetable world we find it abundant. Air-dried wood contains 40 per cent. of H_2O; bread is 37 per cent. water; and of the potatoes and turnips cooked for our dinner, it comprises 75 per cent. of one and 91 of the other. The following table shows the proportion in common vegetables, fruits, and meats:

Mutton	.76	Oysters	.90	Beets	.90
Beef	.72	Salmon	.74	Cabbage	.80
Veal	.78	Apples	.85	Cucumbers	.96
Pork	.72	Carrots	.88	Melons	.90
Eggs	.74				

Water in the Mineral World.—Bodies in which the water is chemically combined in definite proportions, are often called *hydrates.* In the image which the Italian peddler carries through our streets for sale, there is nearly one pound of H_2O to every four pounds of plaster of Paris. One third of the weight of any ordinary soil is this same liquid. In some bodies which are capable of crystallizing, it seems to determine the form and general appearance, and is called "the water of crystallization." If we heat blue vitriol, its water of crystallization will be driven off, and it will lose its color and become white like flour.* A few drops of H_2O will restore the blue. If we expel the

---

* This may be easily shown by filling the bowl of a clay tobacco-pipe with crystals of the salt, and heating them over a lamp or in the fire until the water of crystallization is expelled. Alum may be made anhydrous in the same way.

water from alum, it will puff up, and the transparent crystals will dry into an incoherent mass. Many salts *effloresce, i.e.*, part with their water of crystallization on exposure to the air, and crumble into a white powder.

Water as a Solvent. — Water, having no taste, color, or odor itself, is perfectly adapted to be the general solvent. It becomes at pleasure sweet, sour, salt, bitter, nauseous, and even poisonous. Had water any taste, the whole art of cookery would be changed, since each substance would partake of the one universal watery flavor.

Pure Water. — Rain-water, caught after the air is thoroughly cleansed by previous showers, and at a distance from the smoke of cities, is the purest natural water known. It is tasteless, yet its insipidity makes it seem to us very ill-flavored indeed. We have become so accustomed to the taste of the impurities in water, that they have become to us tests of its sweetness and pleasantness.

River-Water, though it may have less mineral matter than spring-water, is often unfitted for drinking on account of the organic matter it contains. Happily, running water has in itself a certain purifying power, owing to the air which it holds in solution; so that organic substances are burned in it as certainly as they would be in a stove. Still, in order to avoid any danger, river-water should be filtered through charcoal or sand before using.*

---

* A weak solution of potassium permanganate is an excellent test of the presence of organic matter. Place the water to be examined in a glass, and add a few drops of sulphuric acid and a little permanganate; if organic matter is present, the violet permanganate solution is decolorized.

Hard Water.—As water percolates through the soil into our wells, it dissolves the various mineral matters characteristic of the locality.* The most abundant of these are lime, salt, and magnesia. The former produces a *fur* or coating on the bottom of our tea-kettles, if we live in a limestone region. When we put soap in such water, it curdles—*i. e.*, it unites with the lime (CaO), forming a new, or lime soap, which is insoluble in H_2O. H_2O containing an excess of mineral matter, is unwholesome; yet it is probable that the sparkling hard waters of the limestone districts are relished, not only because they are pleasant to the eye and agreeable to the taste, but on account of some hygienic properties in the excess of CO_2 they contain, and possibly because the CaO acts medicinally on the system.†

It is a fact worthy of note that lime and oxide of iron, which are frequently found in H_2O, the latter generally in minute quantities, are both healthful; while the oxides of the other metals are poisonous. Were zinc or barium, for instance, as common

* Most of the water in our world is unwholesome for drinking purposes. On the great ocean, whose volume is more than thirty times that of all the land above sea-level, there is

" Water, water every-where
And not a drop to drink."

† The French authorities are so well satisfied of the superiority of hard water, that they pass by that of the sandy plains, near Paris, and go far away to the chalk hills of Champagne, where they find water even harder than that of London; giving as a reason for the preference that more of the conscripts from the soft-water districts are rejected on account of the want of strength of muscle, than from the hard-water districts. They conclude that calcareous matter is favorable to the formation of the tissues. No positive decision on this point is possible.

near our homes as iron or calcium, wholesome drinking water would be rarely, if ever, found.

Sea-Water.—The most abundant mineral in the
ocean is common salt. Yet sea-water contains traces
of every substance soluble in water, which has been
washed into the sea from the surface of the continents during all the ages of the past. Its saline
constituents are now in the proportion of about one
part in twenty-eight. This amount may be slowly
increasing, as the water which evaporates from the
surface is pure. In this way, the water of the Salt
Lake has become a strong brine, more than one fifth
of its whole weight consisting of saline matter. This
condition would soon disappear if an outlet were
provided.

Water Atmosphere.—As the world of waters is
inhabited, it also has its atmosphere.* Inasmuch as
the H_2O dilutes the O in part, it does not need so
much N as the common air. It is accordingly composed of over one third O instead of only one fifth.
The air, so rich in O, thus absorbed by the water,
gives to it life and briskness. If it be expelled by
boiling, the water tastes flat and insipid.

Paradoxes of Water.—"Cold contracts," is the law
of physics; but as H_2O cools, it obeys this general law
only as far as 39° F. (or 4° C.). Then it slowly expands, cooling down to 32°, its freezing point, when
its crystals suddenly dart out at angles to each other,

---

* Fish inhale O through the fine silky filaments of their gills. When a
fish is drawn out of H_2O, these dry up, and it is unable to breathe, although
it is in a more plentiful atmosphere than it is accustomed to enjoy.

and thus, increasing in size about one twelfth, it congeals to ice. Ice is therefore lighter than water, and so swims on top; otherwise, in severe winters, our northern rivers would freeze solid, killing the fish and aquatic plants. The longest summer could not melt such an immense mass of ice. But now the blanket that Nature weaves over the rivers and ponds prevents the water beneath from reaching the freezing point. We give to water such contradictory terms as "hard" and "soft," "fresh" and "salt." H_2O seems the most yielding of substances, yet the swimmer who falls on his face, instead of striking head foremost, appreciates the mistake, and we could drive a nail into a solid cube of steel almost as easily as into a hollow one perfectly filled with H_2O. H is the lightest substance known, and O is an invisible gas; yet they unite and form a liquid whose weight we have often experienced, and a solid which makes a pavement hard like granite. H burns readily and, when mixed with O, explodes most fearfully; O supports combustion brilliantly—yet the two combined are used to extinguish fires. H or O in excess would destroy life; H_2O is so essential to it that thirst causes a lingering, painful death.

Uses of Water.—The uses of H_2O are as diverse as they are practical. Its properties fit it for a wonderful variety of operations in nature. Its office is not merely to moisten our lips on a hot day, to make a cup of coffee, to lay the dust in the street, and to sprinkle our gardens; it has grander and more profound uses than any of these. *Water is the common* *carrier of creation.* It dissolves the elements of the soil and, climbing as sap up through the delicate capillary tubes of the plant, furnishes the leaf with the materials of its growth. It flows through the body as blood, floating to every part of the system the life-sustaining O, and the food necessary for repairs and for building up the various parts of the "house we live in." It comes from the clouds as rain, bringing to us warmth from the ocean and tempering our northern climate, while in spring it floats the ice of our rivers and lakes away to warmer seas to be melted. It washes down the mountain side, leveling its lofty summit and bearing mineral matter to fertilize the valley beneath. It propels water-wheels, working forges and mills, and thus becomes the grand motive-power of the arts and manufactures. It flows to the sea, bearing on its bosom ships conducting the commerce of the world. It passes through the arid sands, and the desert forthwith buds and blossoms as the rose. It limits the bounds of fertility, decides the founding of cities, and directs the flow of trade and wealth.

PRACTICAL QUESTIONS.

1. Why, in filling the hydrogen gun, do we use 5 parts of common air to 2 of H, and only 1 part of O to 2 of H?

2. Why are coal cinders often moistened with H_2O before using?

3. What injury may be done by throwing a small quantity of H_2O on a fire?

4. Why does the hardness of water vary in different localities?

5. What causes the variety of minerals in the ocean? Is the quantity increasing?

6. Is there not a compensation in the sea-plants, fish, etc., which are washed back on the land?

7. Since "all the rivers flow to the sea," why is it not full?

8. What is the cause of the tonic influence of the sea-breeze?

9. When fish are taken out of the water and thus brought into a more abundant atmosphere, why do they die?

10. Do all fish die when brought on land?

11. What weight of water is there in a hundredweight of sodium sulphate (Na_2SO_4, $10H_2O$), or Glauber's salt?

12. What weight of water in a ton of alum ($KAl2SO_4$, $12H_2O$)?

13. How does the air purify running water?

14. What is the action of potassium permanganate as a disinfectant?

15. What weight of H can be obtained from a liter of water?

16. How much Zn must be employed to obtain 100 grams of H from H_2SO_4?

17. A liter of H under standard conditions weighs 0.0896 gram. What volume of H at 10° C. and 738 *mm.* can be obtained from H_2SO_4 by the action of 8 kilos of Zn?

18. How much $KClO_3$ would be required to evolve sufficient O to burn the H produced by the decomposition of 2 grams of H_2O?

19. How much O would be required to oxidize the metallic Cu which could be reduced from its oxide by passing over it, when white-hot, 20 grams of H gas?

20. How much O would be required to oxidize the metallic Fe which could be reduced in the same manner by 10 grams of H gas?

21. Why are rose-balloons so buoyant?

22. How much H must be burned to produce a ton of water?

CARBON.

Symbol, C. Atomic Weight, 12. Specific Gravity of Diamond, 3.5 to 3.6.

Occurrence.—C is one of the most abundant substances in nature, forming nearly one half of the entire vegetable kingdom, and being a prominent constituent of lime-stone, corals, marble, magnesian rocks, etc. We find it uncombined in two distinct forms or allotropic conditions—viz., the *diamond* and *graphite*.

The Diamond is *pure carbon* crystallized. It is the hardest of all known substances, scratches all

other minerals and gems, and can be cut only by
its own dust. It is infusible, but will burn at a high
temperature. The celebrated diamond beds are in
the East Indies, Borneo, Sumatra, Brazil, Australia,
Mexico, and at the Cape of Good Hope. In 1858,
Brazil furnished 120,000 carats.* They usually appear as semi-transparent, rounded pebbles, inclosed
in a thin, brownish, opaque crust, which being broken,
reveals the brilliant gem within. They are of various
tints, though often colorless and perfectly transparent.
The last are most highly esteemed and, from their
resemblance to a drop of clear spring-water, are
called diamonds of the "first water." They are exceedingly brittle, and valuable gems are said to have
been broken by simply falling to the floor. Nothing
definite is known concerning the origin of this gem.†

---

* A carat is a little less than 4 grains Troy. The term is derived from
the name of a bean, which, when dried, was formerly used in weighing by
the diamond merchants in India.

† Although the diamond is simply pure carbon crystallized, all attempts
to make it have been until recently unsuccessful. A few years ago, it was
discovered that artificial diamonds having all the characteristics of the natural stones could be formed ; but thus far, all that have been made are very
minute and interesting only from a scientific point of view. The value of
the diamond varies with the market; the *general* rule is as follows: a gem
ready for setting, of one carat weight, is worth $150 to $180; beyond this
size, the estimated value increases according to the square of the weight,
but in case of large stones is generally much less than that amount,
although rare beauty or size may greatly enhance the price. The *Kohi-
noor* (mountain of light, now among the crown jewels of England) weighs
103 carats, yet is valued at $10,000,000. Owing to the discovery of many
large diamonds in South Africa, the value of such stones has much decreased
of late. The smaller ones, however, are becoming more expensive on
account of the greater demand for them. The South African diamonds are
seldom colorless, having generally a yellowish tint. Paste diamonds are
now made in Paris, which are so perfect an imitation that only experts
can distinguish them from the real gems.

The Diamond is ground by means of its own powder. Being fitted to the end of a stick or handle, it is pressed down firmly against the face of a rapidly revolving wheel, covered with diamond-dust and oil. This, by its friction, removes the exposed edge and forms a *facet* of the gem. There are three forms of cutting—the *brilliant*, the *rose*, and the *table*. The brilliant has a flat surface on the top, with facets at the side, and also below, the latter terminating in a point, so arranged as to refract the light most brilliantly. This form shows the gem to the best advantage, but is used only in large, thick stones, as it sacrifices nearly half the weight in cutting. The rose is flat beneath, while the upper surface is ground into triangular facets, terminating at a common vertex. The table form is employed for thin specimens, which are merely ornamented by small facets on the edge. The diamond is valued not alone for its rarity and high refractive power, by which it flashes such vivid and brilliant colors, but also for its mechanical uses. For cutting glass, the curved edges of the natural crystal are used.

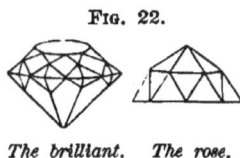

FIG. 22.

The brilliant. The rose.

Graphite or Plumbago, is also called black-lead, because on paper it makes a shining mark like lead. It is found at Ticonderoga, N. Y., Brandon, Vt., and Sturbridge, Mass. It is supposed to be of vegetable origin.

Uses.—Its chief use is for pencils. For this purpose a mixture of powdered graphite and carefully

washed clay is employed.* Though graphite seems
very soft, yet its particles are extremely hard and
the saws used in cutting it soon wear out. We notice this property in sharpening a pencil with a
knife. Graphite mixed with clay is made into black-
lead' crucibles. These are the most *refractory* known
and are used for melting gold and silver. It is also
sold as "British luster," "carburet of iron," "stove
polish," etc., which are employed for blacking stoves
and protecting iron from rusting.

Amorphous Carbon.—This name is given to all
modifications of carbon which are not diamond or
graphite. The term amorphous means without crystalline form. Under this head are included lamp-
black, charcoal, coke, gas-carbon, animal charcoal,
and coal.

Lamp-black is obtained by imperfectly burning
pitch or tar. The dense cloud of smoke is conducted
into a chamber lined with sacking, upon which the
soot collects. It is largely used in painting. It is
mixed with clay to form black drawing crayons, and
with linseed oil to make printers' ink. Lamp-black
has peculiar properties which fit it for printing.
Nothing in nature could supply its place. No matter how finely it is pulverized, it retains its dead-
black color. The minutest particle is as black as the
largest mass. It is insoluble in all liquids. It never

---

* The graphite and clay are mixed with water to a semi-solid mass,
which is then placed in a short iron cylinder having a small opening in the
bottom and forced through the hole by pressure. In this way a long
plastic thread is obtained, which is cut into the required lengths and
heated.

decays. The paper may molder; we may even burn it, and still, in the ashes, we can trace the form of the printed letter. The ancients used an ink composed of gum-water and lamp-black, and manuscripts have been exhumed from the ruins of Pompeii and Herculaneum which are yet perfectly legible.

Soot is unburnt carbon which passes off from a lamp or fire when there is not enough O present to combine with all the C of the fuel. This, therefore, comes away in flakes, and blackens the chimney of the lamp or lodges in the chimney of the house. After a time, a large quantity having collected, we are startled by the cry, "The chimney is on fire!" while with a great roar and flame the soot burns out. This unpleasant occurrence is much more frequent when green wood is used for fuel. The H_2O of the wood absorbs much of the heat of the fire, and so permits the C to pass off unconsumed.

Charcoal is made by burning piles of wood, so covered over with turf as to prevent free access of air. The volatile gases, water, etc., are driven off and the C left behind. This forms about $\frac{3}{4}$ of the bulk of the wood and $\frac{1}{2}$ its weight. Charcoal for gunpowder and for medicinal purposes is prepared by heating willow or poplar wood in iron retorts.

Coke is obtained by distilling the water, tar, and volatile gases from bituminous coal. It is burned in locomotives, blast-furnaces, etc.

Gas-carbon is formed on the interior of the retorts used in coal-gas works. It has a metallic luster, and will scratch glass.

FIG. 23.

Making charcoal.

Animal Charcoal, or bone-black, is made by heating bones in close vessels. Mixed with H_2SO_4, it forms the basis of paste-blacking. It is largely used by sugar-refiners (p. 218). Common vinegar filtered through it becomes the white vinegar of the pickle manufacturers.

Mineral Coal.—This was formed at an early period of the world's history, called the Carboniferous Age. The earth's surface was then pervaded by a genial tropical climate. The air was denser and richer with vegetable food than now. The surface itself was a swamp, moist and hot, in which simple ferns towered into trunks a foot and a half in diameter; and where plants like those which creep at our feet to-day, or are known only as rushes or grasses, grew to the height of lofty trees. The song of bird or hum of insect rarely echoed through the mighty fern-forests; but a strange and grotesque vegetation flourished with more than tropical luxuriance. In these swamps accumulated a vast deposit of leaves and fallen trunks which, under the water, was gradually decomposed. In the process of time the earth settled at various points and floods poured in, bringing sand, pebbles, clay, and mud, filling up all the spaces between the trees that were standing and even the hollow trunks themselves. The pressure of this soil and possibly the internal heat of the earth combined to expel the gases from the vegetable deposits and convert these into mineral coal.* In time, this section was elevated again and another forest flourished, to be in its turn converted into coal. Each of these alternate elevations and depressions produced a layer of coal or of soil. In these beds of coal we now find the trunks of trees, the outlines of trailing vines, the stems and leaves of plants as perfectly preserved as in a herbarium, so that the flora of the Carboniferous Age is nearly as complete as that of our own.

Peat is an accumulation of half decomposed vegetable matter in swampy places.† It is produced

* Where this process was nearly complete, anthracite coal, and where only partially finished, bituminous coal, was formed. The greater the pressure, the harder and purer the carbon produced; unless, however, the covering was not sufficiently porous to allow the gases to escape, when bituminous coal was the result.

† These peat-beds are of vast extent. One tenth of Ireland is covered by them. A bed near the mouth of the River Loire, is said to be fifty leagues in circumference.

mainly by a kind of moss which gradually dies below as it grows above and thus forms beds of great
thickness. Sometimes, however, plants may grow
in the form of a turf and decay, thus collecting a
vast amount of vegetable *débris*. This gradually undergoes a change and becomes a brownish black
substance, loose and friable in its texture, resembling
coal, but, unlike it, containing 20 to 30 per cent.
of O. Peat is used in large quantities as a fuel.
For this purpose, it is cut out in square blocks
and dried in the sun. In some beds it is first finely
pulverized, then pressed into a very compact form
like brick.

Muck is an impure kind of peat, not so fully car-
bonized; though the term is frequently applied to
any black swampy soil which contains a large quantity of decaying vegetable matter. It is used as a
fertilizer.

Various Forms and Uses of Carbon.—We have
seen in what contrary forms C presents itself. It is
soft enough for the pencil-sketch and hard enough
for the glazier's use. Black and opaque, it expresses
thought on the printed page; clear and brilliant, it
gleams and flashes in the diadem of a king. Lamp-
black and charcoal are readily kindled; graphite
resists the heat of the fiercest flame. In the diamond, carbon is an insulator; while in the form
of gas-carbon, it is a conductor of electricity, and
is used in electrical batteries and in the arc light.
In our lamps it gives us light; we burn it in
our stoves and it gives us heat; we burn it in

our engines and it gives us power; we burn it in our bodies and it gives us strength. As fuel, it readily unites with O, yet we spread it as stove-polish on our iron-ware to keep the metal from rusting. It gives firmness to the tree and consistency to our flesh. It is the valuable element of all fuel, burning oils, and gases. Thus it supplies our wants in the most diverse manner, illustrating in every phase the forethought of that Being who fitted up this world as a home for His children. Infinite Wisdom alone would have stored up such supplies of fuel and light and hidden them far under the earth away from all danger of accidental combustion, or anticipated the requirements alike of luxury and the arts.

Properties of Carbon.—Each of these various substances possesses different properties, and yet all have certain properties in common which prove them to consist of one and the same element. They are all tasteless and without color. They are all infusible, and all of them, when heated in air or O, unite with the same proportion of O, forming precisely the same compound—carbon dioxide—from which the C can be obtained again in the form of charcoal. Carbon is the most unchangeable of all the elements, so that even in the charcoal we can trace all the delicate structure of the plant from which it was made. Neither air nor moisture affects it. Wheat has been found in the ruins of Herculaneum that was charred 1800 years ago, and yet the kernels are as perfect as if grown last harvest.

The ground ends of posts are rendered durable by charring. Indeed, some were dug up not long since in the bed of the Thames which were placed there by the ancient Britons to oppose the passage of Julius Cæsar and his army. A cubic inch of fine charcoal has, it is said, 100 feet of surface, so full is it of minute pores. These absorb and condense gases to an almost incredible extent. A bit of C will take up ninety times its bulk of ammonia. As the various gases and the O of the air are brought so closely together within its pores, rapid oxidation is produced, as in the case of spongy platinum (see p. 42). Pans of charcoal soon purify the offensive air of a hospital. Foul water filtered through C loses its impurities. Beer by this process parts not only with its color, but with its bitter taste. Ink is robbed of its value, and comes out clear and transparent as water.

Deoxidizing or Reducing Action of C.—At a high temperature the attraction of C for O is powerful. In the heat of a furnace it will take it from almost the stablest compounds. This fact gives to charcoal great value in the arts. Nearly all the metals and many of the other elements are locked up in the rocks with O, and C is the key made by the Creator for unlocking the treasure-houses of nature for the supply of our wants. By noticing the process of preparing zinc, iron, phosphorus, etc., we shall see the importance of this property of C. A very pretty illustration is shown by placing a few grains of litharge (PbO) on a flat piece of charcoal, and directing upon it the flame of a blow-pipe. The metal will immediately appear in little sparkling globules.

FIG. 24.

PbO *on charcoal.*

Compounds. — Carbon Dioxide, CO_2.—Occurrence.—This gas is commonly known as Carbonic Acid. It is found combined with lime in a large class of salts, known as the carbonates, viz., limestone, marble, chalk, etc., forming nearly one half of their weight, and almost one seventh of the crust of the earth. It comprises $\frac{4}{10000}$ of the atmosphere. It is produced throughout nature in immense quantities. Wherever C burns, in fires, lights, decay, volcanoes—in a word, in all those various forms of combustion of which we spoke under the subject of O, where that gas unites with C, CO_2 is the result. Each adult exhales daily about $8\frac{3}{4}$ oz. of carbon changed to this invisible gas. Each bushel of charcoal, in burning, produces not far from 2500 gallons. A lighted candle gives off about four gallons per hour.

Preparation.—For experimental purposes, CO_2 is prepared by pouring hydrochloric (muriatic) acid on marble or chalk. The reaction may be represented as follows:

$$CaCO_3 + 2HCl = CaCl_2 + H_2O + CO_2.$$

The CO_2 is liberated rapidly and, as it is much heavier than air, may be collected by downward displacement (see Fig. 25), while the calcium chloride remains dissolved in the water of the flask.

The test for CO_2 is clear lime-water. If we expose a saucer of lime-water to the air, the surface of the solution will soon be covered with a thin

FIG. 25.

Preparing CO_2. *

film of calcium carbonate (carbonate of lime), thus showing that there is CO_2 in the atmosphere; or if we breathe by means of a tube through lime-water, the solution will become turbid and milky, thus proving the presence of CO_2 in our breath; by breathing through the liquid a little longer it will become clear, as the carbonate will dissolve in an excess of CO_2.†

* Twist a wire around the neck of a small, wide-mouthed vial, to serve as a bucket. Dip the CO_2 with it *upward* from the jar and test with a lighted match. Dip the H (Fig. 15) *downward*, and test in same way. This illustrates in a striking manner the difference between the gases in respect to specific gravity and combustion.

† Burn a piece of charcoal or a candle in a jar of O. Pour in a little lime-water and shake it well, when there will be a precipitation of chalk (calcium carbonate). Hold a jar of air over a burning lamp or jet of coal-gas, or breathe into the jar and apply the test.

Properties. — CO_2 is a colorless, odorless, transparent gas, with a slightly acid taste, and is a nonsupporter of combustion. Since it is heavier than air, many amusing experiments can be performed

FIG. 26.

Pouring CO_2 down an inclined plane.

with it. It will run down an inclined plane, can be poured from one dish to another, drawn off by a siphon, dipped up with a bucket like water, or weighed in a pair of scales like lead.

To show the C **in** CO_2, hold a strip of Mg foil in a flame until well ignited, then insert in a jar of the gas. White flakes of magnesium oxide* (MgO) mixed with black particles of charcoal will be deposited.

* These may be dissolved by dilute HNO_3, and the black C made more distinct.

Fig. 27.

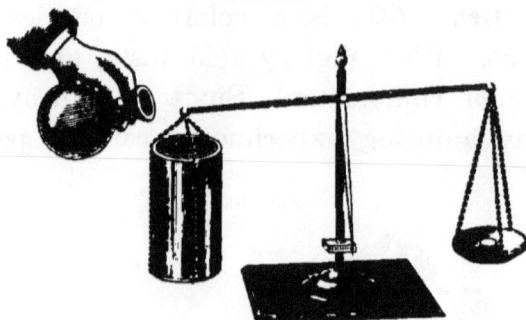

Weighing CO_2.

Asphyxia.—CO_2 accumulates in old wells and cellars, where it has cost the lives of many incautious persons.* The test of lowering a lighted candle should always be employed. If that be extinguished, your life would be in danger of "going out" in the same way, should you descend. The gas may be dipped out like water, or the well may be purified by lowering pans of slaked lime or lighted coals which, when cool, will absorb the noxious gas. The coals may be re-ignited, and lowered repeatedly until the result is reached.† Persons have been suffocated by burning charcoal in an open furnace in a closed room.‡ In France, suicide is sometimes committed in this manner. The

Fig. 28.

Pouring CO_2 *on a light.*

* "Three or four per cent. of CO_2 in the air acts as a narcotic poison by preventing the proper action of the air upon the blood."—MILLER.

† A well in which a candle would not burn within twenty-six feet of the bottom, was thus purified in a single afternoon.

‡ The fumes of burning charcoal owe their deadly property largely to the presence of CO (page 70), one per cent. of which in the air causes headache.

antidote is to bring the sufferer into the fresh air and dash cold water upon his face. In the celebrated *Grotto del Cane*, near Naples, the gas accumulates upon the floor, so that a man living near amuses visitors, for a small fee, by leading his dog into the cave. He experiences no ill effects himself, but the dog falls senseless. On being drawn into the open air, the animal soon revives, and is ready to pick up his bit of black bread and enjoy this reward for his demonstration of the properties of CO_2.

CO_2 in Mines.—Miners call CO_2 *choke-damp*. It is produced by the combustion of *fire-damp* (see p. 71), which accumulates in deep mines,* and when mixed with air, explodes like gunpowder, forming dense volumes of CO_2, which instantly destroys the lives of all who may have escaped the flames of the explosion.† CO_2 has been used for the purpose of extinguishing fires in coal-mines. A mine near Sterling, England, had burned for thirty years, consuming a

---

* The word *gas* was first used in the seventeenth century. Explosions, strange noises and lurid flames had been seen in mines, caves, etc. The alchemists, whose earthen vessels often exploded with terrific violence, commenced their experiments with prayer and placed on their crucibles the sign of the cross—hence the name crucible from *crux* (gen. *crucis*), a cross. All these manifestations were supposed to be the work of invisible spirits, to whom the name *gāst* or *geist*, a ghost or spirit, was applied. The miners were in special danger from these unseen adversaries, and it is said that their church service contained the petition, "From spirits, good Lord, deliver us!" The names "spirits of wine," "spirits of niter," etc., are a relic of the superstitions of that time.

† Where CO_2 alone is found, it is not considered as dangerous as the *fire-damp*, since it will not burn; and it is said that miners will even venture "where the air is so foul that the candles go out, and are then re-lighted from the coal on the wick by swinging them quickly through the air, when they burn a little while and then go out, and are re-lighted in the same way."

seam of coal nine feet thick, over an area of twenty-six acres. CO_2, eight million cubic feet of which were required, was poured into the mine, in a continuous stream, day and night, for three weeks. The mine was then cooled with water, and within a month from the commencement of the operation, was ready for the resumption of work.

Absorption of CO_2 by Liquids. — Water dissolves its own volume of CO_2 under the ordinary pressure of the atmosphere, forming a solution of carbonic acid ; $CO_2 + H_2O$ becoming H_2CO_3. With increased pressure a much greater amount will be absorbed. "Soda water" contains no soda, but is simply H_2O saturated with CO_2 in a copper receiver strong enough to resist the pressure of ten or twelve atmospheres. The gas gives the H_2O a pleasant, pungent, slightly acid taste, and by its escape, when exposed to the air, produces a brisk effervescence.* In beer, ginger-pop, cider, wine, etc., the CO_2 is produced by fermentation.† The gas escapes rapidly through cider and wine, and so produces only a sparkling ; while in a thick, viscid liquid, like beer, the bubbles are partly confined, and hence cause it to foam and froth. In canned fruits, catsup, etc., the "souring" of the vegetables produces CO_2, which sometimes drives out the cork or bursts the bottles with a loud report.

---

* Pass a current of CO_2 through a gill of water. Add a few drops of blue litmus-solution. It will immediately redden. Boil the water, when the gas will escape and the water become blue.

† Dissolve an ounce of sugar in ten times its weight of water. Put it in a flask and add a little fresh brewer's yeast. If kept warm, in a short time it will give off CO_2, which may be tested.

Liquid CO₂.—By a pressure of thirty-six atmospheres, at a temperature of 32° F., CO_2 becomes a colorless liquid very much like H_2O. When this is exposed to the air, it evaporates so rapidly that a portion is frozen into a snowy solid which blisters the flesh like red-hot iron. By means of solid CO_2, Hg can be readily frozen. When mixed with ether and evaporated under the exhausted receiver of an air-pump, a cold of −110° C. may be produced. (See " Physics," p. 191.)

Ventilation.—The relation of CO_2 to life is most important, and can not be too often dwelt upon. We exhale constantly this dangerous gas, and if fresh air is not furnished continuously, we are forced to rebreathe that which our lungs have just expelled.* The languor and sleepiness we feel in a crowded assembly, are the natural effects of the vitiated atmosphere.† The idea of drinking in at every breath the exhalations that load the air of a crowded, promiscuous assembly-room, is a most disgusting one. We shun impurity in every form ; we dislike to wear the clothes of another, or to eat from the same dish ; we shrink from contact with the filthy, and yet sitting in the same room inhale their pol-

---

* It is a fact, as poetical as it is characteristic, that when the air comes forth from the lungs, it is charged with the seeds of disease ; yet, as it passes out, it produces all the tones of the human voice, all songs, and prayers, and social converse. Thus the gross and deadly is, by a divine simplicity, made refined and spiritual, and caused to minister to our highest happiness and welfare.

† It should be noted that the deleterious effects of ill ventilation arise not only from the presence of CO_4, but from the organic particles given off in the breath and exhaled from the skin. (See "Physiology, p. 93.) Re-breathed air is a fruitful source of consumption and scrofula.

FIG. 29.

Testing the currents of air to and from flame.

luted breath. Health and cleanliness alike require
that we should carefully ventilate public buildings,
school-rooms and sleeping apartments.*

Carbon Monoxide, CO, is a colorless, almost odorless gas nearly insoluble in water. It burns with a

* Two openings are necessary to ventilate a room. To illustrate this,
set a lighted candle in a plate of water, as shown in Fig. 29. Cover it with
an open jar, over the neck of which is placed a common lamp-chimney.
The light will soon be extinguished on account of the consumption of O,
and the formation of CO_2. Raise the jar at one side a trifle above the
water, and the candle, if re-lighted, will burn steadily—fresh air coming in
below, and the refuse passing off at the top. Replace the jar, and as the
candle is flickering, insert in the chimney a slip of card, thus dividing the
passage, when the light will brighten again. Hold a bit of smoldering
touch-paper (page 32) at the top, and the smoke will show two opposite
currents of air established in the chimney. Mines have been ventilated in
this way by dividing the shaft. More commonly, however, they have two
shafts at a little distance apart.

pale blue flame, absorbing an atom of O from the air and becoming CO_2. It is seen burning thus in our coal-stoves and at the tops of tall furnace-chimneys. It is often formed abundantly through the action of heated carbon on CO_2. When air enters at the bottom of a clear fire, CO_2 is formed at once; but this gas passing through the hot embers takes up a further quantity of C, becoming changed into CO:* $C + CO_2 = 2CO$, the volume of the gas being exactly doubled in bulk thereby. CO is a deadly poison, and escaping from coal-fires in a close room, has often produced death. Both CO and CO_2 leak through the pores of cast Fe when heated, and still further injure the air of our houses and necessitate ventilation. The offensive odor which comes out on opening the door of our coal-stoves is caused by the compounds of S mixed with the CO.

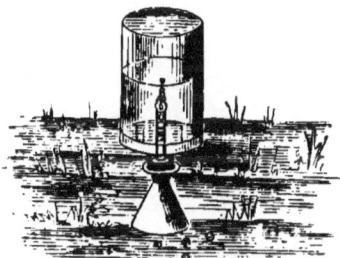

Fig. 30.

Collecting Marsh-gas.

Marsh Gas.—*Light Carburetted Hydrogen*, CH_4. —This we have already spoken of under CO_2, as the dreaded fire-damp of miners. It is colorless, tasteless, odorless, and burns with a pale yellowish flame. It is formed in swamps and low marshy places by the decomposition of

* This fact is of great importance, since thereby much heat is wasted. Stoves are often so constructed as to admit fresh air just above the grate, thus consuming this gas.

vegetable matter, and on stirring the mud beneath, will be seen bubbling up through the water. It may be collected in the manner shown in Fig. 30. It rises from the earth in great quantities at many places. At Fredonia, N. Y., it is used in lighting the village. At Kanawha, Va., it was, until lately, employed as fuel for evaporating the brine in the manufacture of salt. In the oil-wells of Pennsylvania, it frequently bursts forth with explosive violence, throwing the oil high into the air.

Olefiant Gas.—*Heavy Carburetted Hydrogen*, C_2H_4. —This is a colorless gas, with a sweet, pleasant odor, and burns with a clear white light.* It may be easily prepared by heating in a large retort a mixture of one part of alcohol with six of H_2SO_4.

Coal-Gas is a variable mixture of combustible gases and vapors, which may be divided into two classes : 1st. The illuminating constituents, olefiant gas being the most important; and 2d. The diluents, chiefly H, CO, and CH_4. Olefiant gas and hydrocarbons having a similar composition, give whiteness to the flame; while the H, CH_4, and CO have little illuminating power. Bituminous coal is heated in large iron retorts, *B*, until the volatile constituents are driven off and only coke is left. Among the former are coal-tar, NH_3, CO_2, CO, N, compounds of S, CH_4, and C_2H_4.† This mixture is led through the curved

---

* A very characteristic property of olefiant gas is its power of uniting directly with an equal volume of Cl to form a heavy oily liquid called Dutch liquid. It is to this that it owes its name of olefiant gas.

† None of these substances exist in coal. They are formed by the action of heat, which causes the H, C, O, N, and S to combine and make a multiplicity of compounds.

pipes, d, beneath the H_2O in the *hydraulic main*, F; along the tube, g, to the *tar cistern;* thence up and down the *condenser, j.* On the way it becomes cooled and loses its coal-tar, ammoniacal salts,* and liquid hydrocarbons. Lastly, it is passed over lime, $L\ m$, which absorbs the CO_2 and the H_2S.† The remaining gases form the mixture we call "gas." This is collected in the gasometer, P, the weight of which forces it through all the little gas-pipes, and up to every jet in the city.

Coal-gas is very poison-

FIG. 31.

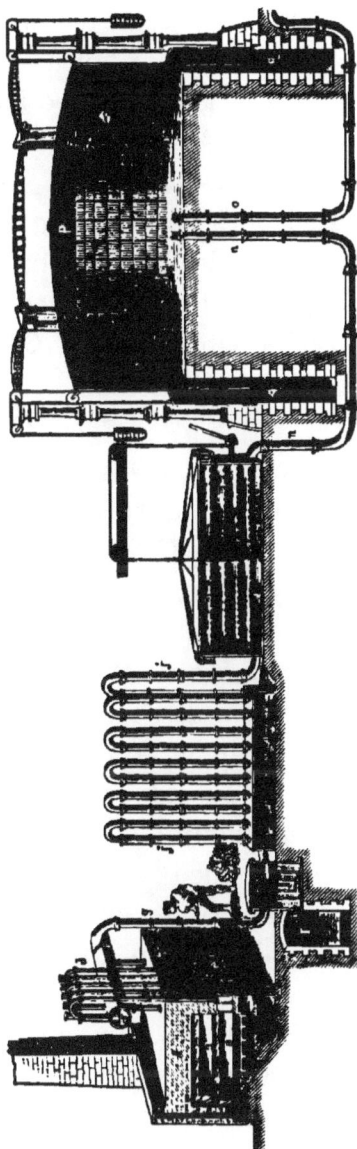

Manufacture of Coal-gas.

* The NH_3 is neutralized by HCl, thus forming chloride of ammonium (sal-ammoniac, NH_4Cl). On evaporation and sublimation, the tough, fibrous crystals of the salt are obtained.

† The removal of the sulphur compounds is especially important, since, when burned, they furnish sulphurous and sulphuric acids. These acids would cause very great injury to books, paintings, and furniture.

ous, and even in small quantities exceedingly delete-rious. When mixed with air, it explodes with great violence. Its unpleasant odor, though often annoy-ing, is a great protection, as we are thereby warned of its presence.

Water-Gas.—This gas, which is now extensively used both for heating and illuminating purposes, is made by leading steam over red-hot coke. The steam (H_2O) is decomposed by the C of the coal with the production of H and CO, as shown in the following equation :

$$C + H_2O = CO + 2H.$$

Both H and CO are gases, and burn with great heat ; when passed through petroleum oils, they become "carburetted" and burn with a luminous flame.

Cyanogen,* $Cy = CN.$—**Preparation.**—As N and C do not combine directly, this gas is obtained in an indirect way. Mix the parings of horns, hides, etc., with pearlash (potassium carbonate) and iron filings, and heat in a close vessel. The N and C of the animal substances, in their nascent state, will com-bine, forming Cy ; this, uniting with the Fe and K, will produce potassium ferro-cyanide (yellow prussiate of potash), a solution of which yields fine yellow crystals. From this salt mercury cyanide is made, which, when heated, decomposes into Hg and Cy.

Properties.—Cy is a transparent, colorless gas, with a penetrating odor. It burns with a characteristic rose-edged purple flame and is exceedingly poisonous.

---

* The term cyanogen means "blue producer "; this gas being the char-acteristic constituent of Prussian blue.

It is very interesting from the fact that, though a compound, it unites directly with the metals like the elements Cl, Br, etc. It is therefore called a *compound radical* (root). We shall find this subject of great importance in Organic Chemistry.

Hydrocyanic Acid, HCy.—Prussic acid, as it is commonly called, is a fearful poison. A single drop on the tongue of a large dog is said to produce instant death. NH_3, cautiously inhaled, is its antidote. Its bitter flavor is detected in peach blossoms, the kernels of plums or peaches, bitter almonds, and the leaves of wild cherry.

Fulminic Acid (*fulmen*, a thunder-bolt).—This compound of Cy is known only as combined with the various metals forming fulminates, which are remarkably explosive. Fulminating mercury was used to fill the bombs with which the life of Napoleon III. was attempted in 1858. It is employed in making gun-caps. A drop of gum is first put in the bottom of the cap, over which is sprinkled a mixture of saltpeter, sulphur, and fulminating mercury, and this is sometimes covered with varnish to protect it from any moisture.

COMBUSTION.

COMBUSTION, in general, is the rapid union of a substance with O, and is accompanied by heat and light.*

Chemistry of a Fire.—Our fuel and lights, such

---

* There are forms of combustion known to the chemist which are not oxidation; as the union of S and Cu. (See page 87, note.)

as wood, coal, oil, tallow, etc., consist mainly of C
and H, and are, therefore, called *hydrocarbons*. In
burning they unite with the O of the air, forming
H_2O and CO_2. These both pass off, the one as a
vapor, the other as a gas. In a long stove-pipe, the
H_2O is sometimes condensed, and drips down, bringing
soot upon our carpets. Ashes comprise the mineral
matter contained in the fuel, united with some of
the CO_2 produced in the fire. When we first put
fuel on a fire, the H is liberated in combination with
some C, in the form of marsh or olefiant gas. This
burns with a flame. Then, the volatile gases having
passed off, we have left the C, which burns without
flame. In maple there is much more C than in pine,
so it forms a good "bed of coals." In the burning
of fuel there is no annihilation; but the H_2O, CO_2,
and the ashes, weigh as much as the wood and the
O that combined with it. No matter how rapidly
the fire burns, even in the blaze of the fiercest
conflagration, the elements unite in exact atomic
proportions.

C is most admirably fitted for fuel, since the
product of its combustion is a gas. Were it a solid,
our fires would be choked, and before each supply
of fresh fuel we should be compelled to remove the
ashes, which would be more bulky than the original
fuel. In the case of a candle or lamp it would be
still more annoying, as the solid product would fall
around our rooms. Still another useful property is
the infusibility of C. Did C melt like Zn or Pb on the
application of heat, how quickly in a hot fire would

the coal and wood run down through the grate and out upon the floor in a liquid mass!

Chemistry of a Candle.—Flame is burning gas. A candle is a small "gas-work," and its flame is the same as that of a "gas-burner." First, we have a little cupful of tallow melted by the heat of the fire above. The ascending currents of cool air which supply the light with O also keep the sides of the cup hard, unless the wind blows the flame downward, when the banks break, there is a *crevasse*, and our "candle runs down." Next, the melted tallow is carried by capillary attraction up the small tubes of the wick into the flame. There it is turned into gas by the heat. Flame is always hollow, and at the center, near the wick, is the gas just formed. If a match be placed across a light, it will burn off at each side, in the ring of the flame, while the center will be unblackened.* The gas may be conducted out of the flame by a small pipe, and burned at a little distance from the candle. Flame is hollow because there is no O at the center. As the gases pass upward and outward from the wick, they come in contact

Fig 32.

Form of flame.

Fig. 33.

Match in flame; the S and P being unconsumed.

---

* Take a sheet of white paper and thrust it quickly down upon the flame of a candle or lamp. It will burn in a ring, and when the paper is removed the center will be found unblackened

with the O of the air, and the H, requiring least heat
to unite, burns first, forming H_2O. The carbon which
was united with the H is now set free in tiny particles, which floating around in the flame of the
burning H become white-hot.* They each send out
a delicate wave of light, and passing on to the outer
part where there is more O, burn, forming CO_2.

FIG. 34.

Testing the CO_2 of a flame by drawing the gas through lime-water.

The flame is blue at the bottom, because there is so
much O at that point that the H and O burn together, and so give little light.

The H_2O may be condensed on any cold surface.
The CO_2 may be tested by passing the invisible
vapor of a candle through lime-water.† The wick of

---

* Frankland has shown that the intensity of a flame, in general, is
determined by the density of the gas: thus, a jet of H burning under a
pressure of ten atmospheres will furnish sufficient light to read a news-
paper at a distance of two feet.

† See also pages 63 and 64.

a candle does not burn, because of the lack of O at the center. It, however, is charred, as all the volatile gas is driven off by the heat. If a portion falls over to the outer part where there is O, it burns as a coal. If we blow out a candle quickly, the gas still passes off and we can relight it with an ignited match held at some distance from the wick. The tapering form of the flame is due to the currents of air that sweep up from all sides toward it. The candle must be snuffed, because the long wick would cool the blaze below the igniting point of C and O, and the C would pass off unconsumed as smoke. A draught of air, or any cold substance thrust into the flame, produces the same result and deposits the C as soot. Plaited wicks are sometimes used, which, being thin, fall over to the outside and burn, requiring no snuffing.

FIG. 35.

Chemistry of a Lamp.—A chimney confines the hot air, and makes a draught of air to feed the flame. A flat wick is used, as it presents more surface to the action of the O.

H_2O *condensed from a flame.*

Argand lamps have a hollow wick which admits air into the center of the blaze. The film which gathers on a chimney when we first light a lamp, is the H_2O produced in the flame condensed on the cold glass. A pint of oil forms a full pint of H_2O. Spirits of turpentine, tar, pine-wood, etc., contain an excess of C, and not enough H to heat it to the igniting point. These, therefore, produce clouds of soot. Alcohol contains an excess of H and little C, hence it gives off great heat and little light.

FIG. 36.

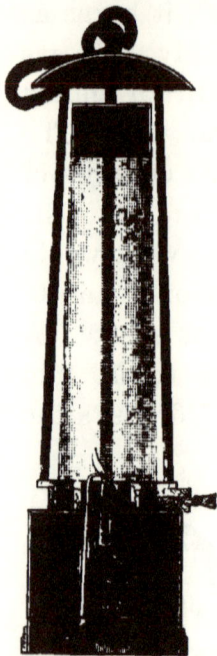

Davy's Safety Lamp.

Davy's Safety Lamp, used by miners, consists of an ordinary oil-lamp surrounded by a cylinder of fine wire gauze. When it is carried into an atmosphere containing the dreaded fire-damp, the flame enlarges and becomes pale, and when the quantity increases, the gas will quietly burn on the inside of the cylinder.* There is no danger of an explosion so long as the gauze remains perfect and draughts are avoided.† Through carelessness, however, fearful accidents have occurred. Miners become extremely negligent, and an account is given of an explosion, in which about a hundred persons were killed, caused by a lamp being hung on a nail by a hole broken through the wire gauze.

FIG. 37.

Wire gauze over flame.

* The principle of the lamp can be illustrated by holding a fine wire gauze over the flame of a candle or gas-burner (Fig. 37). The flame will not pass through, since the wire will conduct away the heat and so reduce the temperature below the igniting point. A jet of gas, issuing at a low temperature, may be lighted on either side of the gauze at pleasure.

† At such a time, however, the wise miner will leave the place of danger, lest the metal should melt and the fire escape to the gas, when an explosion would ensue,

The **Bunsen Burner,** which is used in the laboratory, consists of a gas-jet, a, surrounded by a metal tube, c, at the bottom of which are openings, b, for the admission of air.

The gas passes up the tube, mingles with the air which it draws in through the openings, and burns at the top without smoke. The O is supplied in sufficient quantity to burn the H and C simultaneously; hence there is great heat with little light, and no soot is deposited on the bottom of dishes heated by this burner.

FIG. 38.

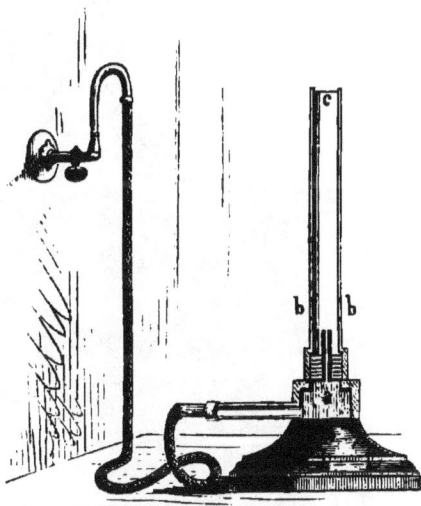

The Bunsen Burner.

The **Oxy-hydrogen Blow-pipe** is so constructed that a jet of O is introduced into the center of one of burning H, thus producing a *solid* flame. A watch-spring will burn in it with a shower of sparks. Pt, the most infusible of metals, will readily melt. In the common flame, as we have seen, the little particles of solid C, heated by the burning H, produce the light. As there is no solid body in the blow-pipe flame, it is scarcely luminous. If, however, we insert in it a bit of CaO, or MgO, a dazzling light is produced. This is called the "Drummond," "Lime," or

FIG. 39.

The Oxy-hydrogen Blow-pipe.

"Calcium" Light, and with a properly arranged reflector has been seen at a distance of one hundred and eight miles.

Mouth Blow-pipe.—In the common blow-pipe, used by jewelers and mineralogists, a current of air from the mouth* is thrown across the light just above the wick. The flame loses its brilliancy and is driven one side in the form of a cone (Fig. 41). Its size, also, is less, and since the combustion is concentrated into a smaller space, its temperature is higher than that of an ordinary flame. The hottest point is at b, a little beyond the tip of the inner blue cone, because at this place the combustion is most complete. The inner cone contains CO in excess, hot and ready to combine with O from any substance exposed to it, and is therefore called the *reducing flame.* The outer envelope contains

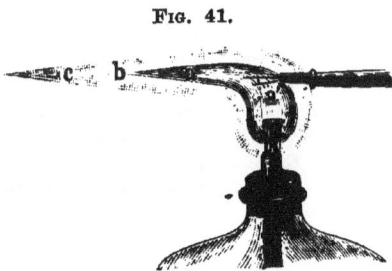

Fig. 40.

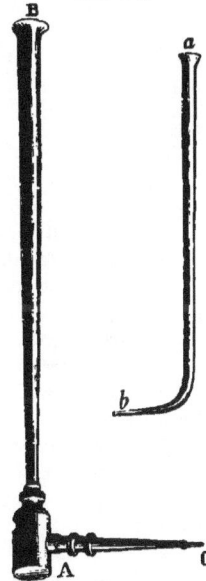

Common Blow-pipes.

Fig. 41.

the O in excess, borne forward by the jet of flame, highly heated by it, and ready to unite with a metallic body. It is therefore called the *oxidizing flame.—Example:* Hold a copper cent in the "reducing flame"; its rust, copper oxide, will be

* The air must come from the *mouth* by the action of the muscles of the cheeks, not from the lungs.

cleaned off, and the metal will shine as brightly as if just from the mint. In the "oxidizing flame" a film of copper oxide will be formed over 'the surface, and as we move the cent the most beautiful play of colors will flash from side to side.*

PRACTICAL QUESTIONS.

1. Why will pine-wood ignite more easily than maple?

2. Why is fire-damp more dangerous than choke-damp?

3. Represent the reaction in making CO_2, showing the atomic weights, as in the preparation of O on page 12.

4. Should one take a light into a room where the gas is escaping?

5. Why does it dull a knife to sharpen a pencil?

6. Where was the C, now contained in the coal, before the Carboniferous age?

7. Must the air have then contained more plant food? (p. 58.)

8. What is the principle of the aquarium?

9. What test should be employed before going down into an old well or cellar?

10. What causes the sparkle of wine, and the foam of beer?

11. What causes the cork to fly out of a catsup bottle?

12. What physical principle does the solidification of CO_2 illustrate?

13. Why does the division in the chimney shown in Fig. 29 produce opposite currents?

14. What causes the unpleasant odor of coal-gas? Is it useful?

15. What causes the sparkling often seen in a gas-light?

16. Why does H in burning give out more heat than C?

17. Why do not stones burn as well as wood?

18. Why does not hemlock make "a good bed of coals"?

19. What adaptation of chemical affinities is shown in a light?

20. Why does snuffing a candle brighten the flame?

21. Why is the flame of a candle red or yellow, and that of a kerosene oil lamp white?

---

* Introduce a small piece of common flint-glass tube into the *reducing flame*. The glass will become opaque and black, because the Pb will be reduced from the transparent form of silicate to the opaque condition of metal. When this has happened, place the black portion just in front of the oxidizing flame. The discoloration will slowly disappear, and the Pb will recombine with O from the air and the glass again become transparent.

22. Why does a street gas-light burn blue on a windy night? Is the light then as intense? The heat?

23. Why does not the lime burn in a calcium-light?

24. Why is a candle-flame tapering?

25. Why does a draught of air cause a light to smoke?

26. What makes the coal at the end of a candle-wick?

27. Which is the hottest part of a flame?

28. Why does not a candle-wick burn?

29. How does a chimney enable us to burn without smoke highly carboniferous substances like oil?

30. How much CO_2 in 200 lbs. of chalk?

31. What weight of CO_2 in a ton of marble?

32. Why does not a cold saucer held over an alcohol flame blacken, as it does over a candle or gas-light?

33. How much CO_2 is formed in the combustion of one ton of C?

34. What weight of C is there in a ton of CO_2?

35. How much O is consumed in burning a ton of C?

36. What weight of sodium carbonate (Na_2CO_3, "carbonate of soda") would be required to evolve 12 grams of CO_2?

37. How much CO_2 will be formed in the combustion of 30 grams of CO?

38. What weight of $CaCO_3$ would be required to evolve 12 grams of CO_2?

39. What would be the volume of these 12 grams of CO_2 at 12° C. and 744 $mm.$?

40. How much C would be necessary to furnish CO_2 enough to fill a gas-holder 10 meters high and 4 meters in diameter when the temperature is 25° C. and the barometer stands at 754 $mm.$?

41. Write in double columns the different properties of carbon dioxide and carbon monoxide; thus,

CO_2 is	CO is
1, non-inflammable.	1, inflammable.

THE ATMOSPHERE.

THE "air we breathe" consists chiefly of N and O, mixed in the proportion of 79 parts of N to 21 of O by volume, or 77 of N to 23 of O by weight. Besides the N and O, air always contains CO_2 and watery vapor, the former amounting in volume to 4 parts in 10,000, and the latter varying in amount.

A very clear idea of the proportion of these several
constituents may be formed by conceiving the air,
not as now dense near the surface of the earth, and
gradually becoming rarefied as we ascend,* but of a
density throughout equal to that which it now possesses near the earth. The atmosphere would then
be about five miles high. The vapor would form
upon the ground a sheet of H_2O five inches deep,
next to this the CO_2 a layer of 13 feet, then the O
a layer of one mile, and last of all the N one of four
miles.†—GRAHAM. In this arrangement we have supposed the gases to be placed in the order of their
specific gravity. The atmosphere is not thus composed in fact, the various gases being equally mingled throughout, in accordance with a principle called
the "*Law of the diffusion of Gases.*" If we throw a
piece of lead into a brook, it will settle instantly to
the bottom by the law of gravitation and will remain
there by the law of inertia. But if we throw into
the atmosphere a quantity of CO_2, it will sink for
an instant, then immediately begin to mingle with
the surrounding air and soon become dissipated.—
Example: If we invert an open-mouthed bottle full
of H over another full of CO_2, the H, light as it is,
will sink down into the lower jar; and the CO_2,
heavy as it is, will rise into the upper jar; and in a
few hours the gases will be found equally mixed.

---

* At a height of about 38 miles the air is only $\frac{1}{100}$ as dense as at the
surface of the earth. At a height of 50 miles it is so extremely rare that
this is usually given as the height of the atmosphere.

† The N and O form so large a part, that they are considered in ordi-
dinary calculation to compose the whole atmosphere.

By this law the proportion of the elements of the atmosphere is the same every-where, and has not varied within historic times. Samples have been analyzed from every conceivable place, from polar and torrid regions, from prairies and mountain-tops, from balloons and mines, from crowded capitals and lonesome forests, and even from bottles found sealed up in the ruins of Herculaneum, and the result is almost exactly the same. These gases do not form a chemical compound, but a mere mechanical mixture,* and they are as distinct in the air as so many grains of wheat and corn mingled in a measure.

Fig. 42.

Diffu-
sion of
gases.

Each of the constituents of the air has its separate use and mission. The action of O and N we have already seen.

Uses of CO_2. — Carbonic acid bears the same relation to vegetable that O does to animal life. The leaf — the plant-lungs — through its million of little *stomata*, or mouths, drinks in the CO_2. In that minute leaf-laboratory, by the action of the sunbeam,

* "To illustrate the difference between a mechanical mixture and a chemical compound, mix powdered S and filings of Cu. The color of the S as well as that of the Cu will disappear, and to the unaided eye will present a uniform greenish tint; with the microscope, however, the particles of Cu may be seen lying by the side of those of S; and we can wash away the lighter S with H_2O, leaving the heavier Cu behind. Here no *chemical action* has occurred; the S and Cu were only *mechanically mixed*. If we next gently heat some of the mixture it soon begins to glow, and on examining the mass we find that both the Cu and the S have disappeared as such, that they can not be distinguished even with the most powerful microscope, and that in their place we have formed a black substance possessing properties entirely different from those possessed either by the Cu or the S."—ROSCOE.

the CO_2 is decomposed,* the C being applied to build
up the plant, and the O returned to the air for our use.
Plants breathe out O as we breathe out CO_2. We
furnish vegetables with air for their use, and they in
turn supply us. There is thus a mutual dependence
between the animal and the vegetable world. Each
relies upon the other. Deprived of plants, we should
soon exhaust the O from the air, supply its place with
CO_2, and die; while they, removed from us, would
soon exhaust the CO_2, and die as certainly. We pol-
lute the air while they purify it. Each tiny leaf
and spire of grass is thus imbibing our foul breath,
and returning it to us pure and fresh.† This inter-

---

* "In order to decompose carbonic acid in our laboratories, we are
obliged to resort to powerful chemical agents, and to conduct the process
in vessels composed of resisting materials, under all the violent manifestations of light and heat, and we then succeed in liberating the carbon only
by shutting up the oxygen in a still stronger prison; but under the, quiet
influences of the sunbeam, and in that most delicate of all structures, a
vegetable cell, the chains which unite together the two elements fall off,
and, while the solid carbon is retained to build up the organic structure,
the oxygen is allowed to return to its home in the atmosphere. There is
not in the whole range of chemistry a process more wonderful than this.
We return to it again and again, with ever increasing wonder and admiration, amazed at the apparent inefficiency of the means, and the stupendous
magnitude of the result. When standing before a grand conflagration, wit-
nessing the display of mighty energies there in action, and seeing the
elements rushing into combination with a force which no human agency
can withstand, does it seem as if any power could undo that work of
destruction, and rebuild those beams and rafters which are disappearing
in the flames? Yet in a few years they will be rebuilt. This mighty force
will be overcome; not, however, as we might expect, amidst the convul-
sion of nature, or the clashing of the elements, but silently, in a delicate
leaf waving in the sunshine."—COOKE.

† From this statement it is evident that the foliage of house-plants
must be healthful. Moreover, there is some reason to believe that the O
which they exhale is highly ozonized, and therefore of great value in
destroying miasmic germs. We should remember, however, that flowers
exhale CO_2; and the odor of certain plants, and the pollen of others, are

FIG. 43.

Apparatus arranged to catch the O evolved from a sprig of leaves.

change of office is so exactly balanced, that, as we have seen, the proportion of CO_2 and of O, in the open air, never varies.*

very injurious. Plants and flowers, which to one person are inocuous, are to another detrimental. Thus the fragrance of new-mown grass, which is so agreeable to some, produces in others what is termed the *hay-fever;* due, it is said, to the pollen of the grass. Each family, therefore, must determine for itself what should be excluded from its collection. It is evident that flowerless plants, like the ivy, etc., are harmless, while the cheerfulness given to an apartment by even a few pots of flowers on a window-bench, should induce one to take some trouble in order to make a selection which will not only beautify but purify the room.

* "Two hundred million tons of coal are now annually burned, producing six hundred million tons of CO_2. A century ago, hardly a fraction of that amount was burned, yet this enormous aggregate has not changed the proportion in the least."—YOUMANS.

Plants Store up Solar Energy. — The sunbeam, which is thus strong enough to wrench apart the C and O, sends out the O full of potential energy and builds up the plant. The energy of the sunbeam is transformed into the potential energy of O and of the vegetable structure. The sun shining on a meadow causes the grass to grow. If the hay made from it be eaten by an animal, the same amount of energy will be liberated as was received from the sun. A tree towers upward through a century of sunshine. When burned, it sets free as much energy as was needed to perfect its growth. A bushel of corn, then, represents not alone so much C, H, and O, but also an amount of sun-energy which is available for any purpose to which we wish to apply it. (See *Conclusion.*)

Animals Spend Solar Energy. — In the process of digestion the energy stored in the plant is transferred to the animal, is given out by its muscles on their oxidation and produces motion, heat, etc. NH_3, CO_2, and H_2O are decomposed by the plant and organized into complex molecules (see p. 185), full of potential energy. The animal oxidizes the organic molecules, and breaks them up into NH_3, CO_2, and H_2O again—simple molecules robbed of energy which the animal has used. Thus the plant builds up and the animal tears down. The plant garners in the sunbeam and the animal scatters it again. The plant reduces and the animal oxidizes.

Uses of Watery Vapor. — We have already seen the uses of H_2O. As vapor, it is every-where present and ready to supply the wants of animals and plants. Were the air perfectly dry, our flesh would become shriveled like a mummy's, and leaves would wither as in an African simoom. Rivers and streams flow to the ocean; yet all their fountains are fed by the currents that move in the air above us. H_2O rises as vapor, flows on to colder regions, falls as rain, sleet, snow, or hail, and then wends its way back to the ocean, turning many a water-wheel on its way as it parts with its potential energy.

Permanence of the Atmosphere.—The elements of the air unite to form HNO_3 only by the passage of electricity, and then in minute quantities. If they combined more readily we should be constantly exposed to a shower of this corrosive acid that would be destructive to all vegetation, clothing, and even our bodies themselves.—O, N, and CO_2 are converted into liquids only by an apparatus specially made for the purpose, and under circumstances which could rarely, if ever, occur in Nature.* These substances are therefore constantly in the condition promptly to supply the demands of animals and plants.—Watery vapor, on the contrary, is deposited as dew or rain by even slight changes of temperature; this readiness of condensation is equally necessary to meet the wants of animal and vegetable life.

---

* The liquefaction of the so-called "permanent gases," N, O, H, etc., was first accomplished in 1877 by Cailletet of Paris and Pictet of Geneva, almost simultaneously.

THE HALOGENS.

Chlorine.. Symbol, Cl; Atomic Weight, 35.5; Specific Gravity, 2.45.
Iodine.... " I; " " 127.; " 4.95.
Bromine.. " Br; " " 80.; " 3.19.
Fluorine.. " F; " " 19.

THESE four elements are closely allied, and are known as the *halogens*, from *hals*, salt, because they form a class of compounds (Haloids) which resemble common salt (NaCl).*

CHLORINE is named from its green color. It is chiefly found in salt, of which it forms 60 per cent. It may be prepared by gently heating a mixture MnO_2 and hydrochloric acid:

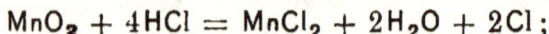

$$MnO_2 + 4HCl = MnCl_2 + 2H_2O + 2Cl;$$

or still more conveniently from NaCl, MnO_2, and H_2SO_4. On slightly warming this mixture a regular and copious evolution of Cl takes place. Cl is heavier than common air, and hence may be collected by displacement, as in the preparation of CO_2.

Properties.—Cl has a greenish-yellow color, and a peculiarly disagreeable odor. It produces a suffo-

---

* In comparing the halogens with one another, the chemical activity of F, which has the smallest atomic weight, is the most powerful; next in the order of activity is Cl, then Br, and, lastly, I, the atomic weight increasing as the chemical energy declines. Cl is gaseous, Br, liquid, and I solid. The specific gravity, the fusing point, and the boiling point, rise as the atomic weight increases. The halogens combine energetically with the metals, and, when united with the same metal, furnish compounds which are *isomorphous*; that is to say, they all crystallize in the same form—potassium fluoride, chloride, bromide, and iodide, for example, all crystallize in cubes. Each, also, forms with H a soluble, powerful acid —HCl, HI, HBr, HF.

Fig. 44.

Preparing Cl.

cating cough, which can be relieved by breathing ammonia or ether. It is incombustible. If a lighted candle is lowered into a jar of Cl, it burns with a red smoky flame for a short time, and then goes out. Arsenic, Dutch gold-leaf, phosphorus, etc., combine with it so rapidly as to inflame. Powdered antimony slowly dropped into it produces a shower of brilliant sparks. Cold water absorbs about twice its volume of the gas.

Cl has a very strong affinity for H. When H and Cl are mixed in the dark and exposed to direct sunlight, they unite with an explosion; and Cl is able to take H away from its combination with other substances. Thus it acts energetically on turpentine ($C_{10}H_{16}$), uniting with its H and setting its C

free in a great cloud of soot. Another example is
seen in the way in which
a candle burns in the
gas. It will even decom-
pose water to get H, set-
ting O free in the process.
This action, like the di-
rect union of H and Cl,
does not take place in
the dark; but if chlorine
water is exposed to direct
sunlight, it goes on slowly
and the O can be col-
lected and tested. The reaction is represented by the
equation :

FIG. 45.   FIG. 46.

Candle in Cl.   *Turpentine in* Cl.

$$H_2O + 2Cl = 2HCl + O.$$

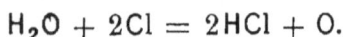

This same reaction also takes place without the
aid of sunlight when there is some substance present
on which the O can readily act. Such substances
are organic coloring matters, and hence Cl can destroy many dye-stuffs and thus bleach cloth, etc., *in
the presence of moisture.* Upon some dyes it probably
acts directly, converting them into colorless substances, but in many cases the color is discharged
only when moisture is present. In such cases the
bleaching is a burning or oxidation, and hence Cl is
called an "indirect oxidizing agent." The necessity
for the presence of moisture is shown by placing two
strips of colored calico or of litmus paper, one dry
the other moistened, in different jars of *dry* Cl. The
O is set free by the Cl in the *nascent state,* and

hence produces an effect of which the atmospheric O is incapable. Cl discharges the color of indigo, writing-ink, wine, etc., almost instantaneously, but has no effect on printer's ink, the basis of which is C, nor on mineral colors in general.

Uses.—Bleaching.—In domestic bleaching the cloth is first boiled with strong soap, to dissolve the grease and wax, and then laid upon the grass, being frequently wet to hasten the action of the air and sun.* It seems likely that ozone is formed by the evaporation of the moisture, and that the bleaching is really an oxidation by its means. Cl is now extensively used for bleaching cloth, paper-pulp, etc.† It is usually employed in the form of so-called "chloride of lime" or bleaching powder.

Disinfectant.—Cl is a powerful disinfectant. It breaks up the offensive substance by uniting with its H as in bleaching. Other disinfectants, as burnt paper, sugar, etc., only disguise the ill odor by substituting a stronger one. In the sick-room Cl is set

* This was essentially the process long pursued in Holland, where linens were formerly carried for bleaching; hence the term "Holland linen," still in use. The H_2O about Haarlem was thought to have peculiar properties, and no other could compete with it. Cloths sent there were kept the entire summer, and were returned in the fall. Later a similar plan was adopted in England. But the vast extent of grass-land required, the time occupied, and the temptation to theft, made the process extremely tedious and expensive. The statute laws of that time abound in penalties for cloth stealing. It is estimated that all the men, women, and children in the world could not, by the old way, bleach all the cloth that is now used.

† Stains can be removed from *uncolored cloth* by "Labarraque's Solution," a compound of Cl, which can be obtained of any druggist. Place the cloth in this liquid, and if the stain is obstinate, pour on a little boiling H_2O, or place it in the sun for some hours. Then rinse thoroughly in cold H_2O and dry.

free from chloride of lime (bleaching powder) by exposing it to the air in a saucer with a little H_2O. The gas gradually passes off, being set free by the action of the CO_2 in the air, and the process may be hastened by adding a few drops of dilute acid. Chloride of lime is, therefore, of great service for disinfecting all places exposed to any noxious or unpleasant effluvia. Hospitals and rooms in which persons have died of a contagious disease are purified by placing in them pans full of a mixture which is disengaging Cl in large quantities.

Compounds.—Hydrochloric Acid, *Muriatic Acid,* HCl.—Cl and H form only this one compound, and it is produced whenever Cl acts upon H or any substance (*e.g.,* turpentine, H_2O) which contains H. If a lighted jet of H be brought into a jar of Cl, the flame becomes whitish, the greenish Cl disappears and a gas is formed which fumes as it escapes from the jar.

HCl is prepared by treating common salt (NaCl) with H_2SO_4:

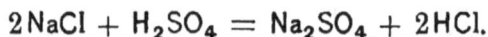

$$2\,NaCl + H_2SO_4 = Na_2SO_4 + 2\,HCl.$$

Properties.—It is an irrespirable, irritating, acid gas, with an intense attraction for H_2O, which causes it to produce white fumes in the air. Water at 60° F. will absorb over 450 times its volume of the gas,* producing the liquid known as "*Hydrochloric*" or

---

* The great solubility of HCl in water can be shown by an experiment like that employed to demonstrate the solubility of NH_3 (see p. 36). If the water be colored with blue litmus, it will turn red as it comes into the bottle.

FIG. 47.

Preparing HCl.

"*Muriatic Acid.*" This dissolves many metals with evolution of H and formation of chlorides. When pure it is colorless, but has ordinarily a yellow tinge, due to various impurities. Its tests are NH_3, with which it forms a white cloud of sal-ammoniac fumes, and silver nitrate, from which it precipitates AgCl. With HNO_3 it makes aqua regia,* or *royal water*, so called because it dissolves Au, the "king of the metals"; Cl is set free, which, in its nascent state, attacks the Au and combines with it.

Acids, Bases, and Salts.—We have seen that when gaseous NH_3 and-HCl come together, they unite with the formation of a white cloud. The same substance is produced when a solution of HCl is poured into a solution of NH_3, and may be obtained as a white

* Boil HCl in a test tube with fragments of gold-leaf. They will not dissolve. Add a few drops of HNO_3, and a yellow solution of gold chloride will be quickly formed.

powder by evaporating off the water. We have also seen that NH_3 or its solution turns red litmus blue, and that HCl or its solution turns blue litmus red. Now, if, before pouring the hydrochloric acid into the NH_3 solution, a little litmus be added to the latter, it will be found that it remains blue until a certain quantity of the acid has been poured in, and that then the color suddenly changes to red. The point at which this change takes place may be hit so exactly that a drop of acid will turn the solution red, or a drop of NH_3 solution blue. At this point the solutions are said to have *neutralized* each other, since the distinctive character of each has disappeared.

These solutions and the product of their action on each other may stand as types of those very important classes of chemical compounds: the acids, the bases, and the salts.

The **Acids** are generally sour* and turn vegetable colors—such as the infusion of blue litmus, or of purple cabbage†—to a bright red. This will be seen to be the case with even very dilute solutions of the three acids with which we have thus far become familiar, HNO_3, H_2SO_4, and HCl. They all have the power of neutralizing NH_3 solution, and substances like it, with the formation of compounds in which

---

* Certain acids, as well as certain bases, are insoluble in water, and hence have no taste. They, however, combine to form salts, which is their true test.

† Paper tinged blue with a solution of litmus (a coloring matter obtained from certain lichens) should be constantly at hand in the laboratory, to determine the presence of a free acid. The same paper faintly reddened by vinegar, or any other acid, is a convenient test for the alkalies. The cabbage solution is made by steeping red cabbage-leaves in water, and straining the purplish liquid thus obtained.

the place of the H of the acid has been taken by a metal or something which acts like a metal. All acids contain H which can be thus replaced.

The **Bases** are substances which, like NH_3 solution, have the power of neutralizing acids. They all contain a metal or something which acts like a metal. *The alkalies** are bases which are soluble in water, have a soapy taste and feel, turn red litmus to blue, and red-cabbage solution to green, neutralize the acids and restore the colors changed by them. The property which the acids and bases thus have of uniting with each other and destroying the chemical activity which either possesses alone is their distinguishing trait.†*

The **Salts** are substances formed by the neutralization of an acid by a base. They may be defined as acids in which the whole or part of the H has been replaced by a metal. Thus we have as salts of HNO_3: KNO_3, $Pb2NO_3$, etc.; of HCl: $NaCl$, $CaCl_2$, etc.; of H_2SO_4: $NaHSO_4$, Na_2SO_4, etc.

Nomenclature of Acids and Salts.—The termination *ic* is generally used in naming acids. Thus

---

* The alkalies are compounds of H, O, and a metal. They are hence called *hydroxides:* as KOH (potassium hydrate, caustic potash), NaOH (sodium hydrate). Ammonia solution is supposed to contain the compound NII₄OH (ammonium hydroxide), in which NH₄ plays the part of a metal.

† To a part of the purple-cabbage solution add a few drops of a solution of caustic potash: a green liquid will be produced. To another portion add a few drops of sulphuric acid: the solution will become red. Pour the red acid liquor into the green alkaline one, and stir the mixture: the red color at first disappears, and the whole remains green; but on adding it cautiously, a point is reached at which it assumes a clear blue color. There is then no excess of acid or alkali; and on evaporation. a neutral salt, *potassium sulphate*, may be obtained.

HCl, hydrochloric acid; HNO_3, nitric acid; H_2SO_4, sulphuric acid. All the acids except some of those in which members of the halogen group appear, contain O, and in some cases these differ from each other only in the proportion of O. Thus, there are two acids containing N: HNO_3, and HNO_2. To distinguish these the latter is called *nitrous* acid; and in general, where there are two similar acids like these, the termination *ous* is employed in naming that containing the less proportion of O. The salts derived from the acids that end in *ic* take the termination *ate*, and those from acids in *ous*, the termination *ite*. Thus from HNO_3 we have *nitrates;* from HNO_2, *nitrites.* An exception to this rule is made in naming the salts of HCl and other halogen acids which contain no oxygen; their salts are *binary** compounds and take the termination *ide.* Thus we have NaCl, sodium chloride, etc.

BROMINE—named from its bad odor—is a poisonous, volatile, deep-red liquid, with the general properties of Cl.† It occurs only in combination, chiefly with Na in NaBr, which is principally found in seawater. Br can be prepared from NaBr in the same way as Cl from NaCl. With H, Br forms hydrobromic acid, HBr, which resembles HCl. With metals, bromides are formed (*e. g.* $CdBr_2$), which are used in photography and in medicine.

---

* See Introduction, page 6.
† Br is the only element, except Hg, which is liquid at ordinary temperatures.

IODINE is named from its beautiful violet-colored vapor. It is made from kelp (the ashes of sea-weed). Its salts are found in sea-water and in some mineral springs. It crystallizes in bluish-black scales, emits a smell resembling that of Cl, sublimes* slowly, and is deposited in crystals on the sides of the bottle in which it is kept. I is sparingly soluble in H_2O, but readily in ether or alcohol. It inflames spontaneously when in contact with phosphorus.† Its compounds with the metals, called the iodides, are remarkable for their variety and brilliancy of color. (See *Appendix.*) It stains cloth a yellowish tint, which may be removed by a solution of potassium iodide. Its test is starch, forming the blue iodide of starch.‡ I reveals the presence of this substance in potatoes, wheat, etc.§ It is much used in medicine to scatter scrofulous or cutaneous eruptions and swellings.

FLUORINE is the only element that will not unite with O. It exists, in small quantities, in the enamel of the teeth. It is found in Derbyshire or fluor spar

* A body is said to *sublime* when it rises as vapor and condenses in the solid form; when it condenses as a liquid it is said to *distil.*

† Place on a clean, white dish a few scales of iodine and a bit of phosphorus as large as a pea. They will soon combine, igniting the phosphorus and subliming a part of the iodine.

‡ Mix one or two drops of a solution of potassium iodide with a little dilute starch mucilage; no change of color will occur. Add a single drop of Cl water to the mixture; an immediate coloration will occur, owing to the combination of the Cl with the K, while I is set free, which acts upon the starch. Add a little more chlorine water; the color disappears, owing to the formation of chlorine iodide, which is without action on starch.

§ Pour a few drops of a solution of iodine in alcohol on a freshly-cut potato. Blue specks will show the presence of starch.

(CaF_2), of which beautiful ornaments are made. Fluorine has not been obtained in the free state. Its affinities are so strong that when freed from one compound it immediately enters into new combination with the substances with which it comes in contact. With H, it forms hydrofluoric acid (HF), noted for its corrosive action on glass. This eats out the silica or sand from the glass, and is therefore used for etching labels on glass bottles and on shop windows.—*Experiment:* Powdered fluor spar is placed in a lead tray, and covered with dilute H_2SO_4. The heat of a lamp applied beneath, for a moment only, liberates the gas in white fumes very rapidly. The plate of glass is covered with wax, and the design to be etched is traced upon it with a sharp-pointed instrument. This is then laid over the tray, and the escaping gas soon etches the lines laid bare into an appearance like ground grass. A solution of HF in H_2O is often sold for this purpose. It is kept in lead or gutta-percha bottles, combines with H_2O with a hissing sound, like red-hot iron, and must be handled with care, as even a minute drop will sometimes produce obstinate blisters and sores.

SULPHUR.

Symbol, S... Atomic Weight, 32... Specific Gravity, 1.98-2.05.

Sources.—S is found native in volcanic regions. It is mined at Mount Ætna in great quantities. United with the metals it forms sulphides, known as cinnabar, iron pyrites, galena, blende, etc. Combined with O and metals it exists in gypsum (plaster), heavy spar, and other sulphates. It is found in the hair, and many dyes contain Pb which unites with the S, and forms a black compound that stains the hair. It is contained in eggs, and so tarnishes our spoons by forming a sulphide of silver. It is always present in the flesh, and hence manifests itself in our perspiration. In commerce it is sold as brimstone, formed by melting S and running it into molds; also as flowers of sulphur, obtained by sublimation.

Properties.—It is insoluble in H_2O, and hence tasteless. Its solvents are carbon disulphide (CS_2), oil of turpentine, and benzol. It is a non-conductor of heat, and crackles when we grasp it with a warm hand. It may be obtained in several allotropic forms: 1st, octahedral crystals; 2d, prismatic crystals; 3d, an amorphous (without form) or uncrystallized state; and 4th, a viscid condition. The last is the most interesting.—*Example:* When S is melted, and then heated more strongly, it changes to a thick, viscid, dark-colored liquid resembling molasses. If this is poured into cold water, it becomes elastic like India rubber. In this form it is used for taking impressions of medals, coins, etc. (See *Appendix*.)

Uses.—On account of its ready inflammability, S is employed in the making of matches and gunpowder, but its chief consumption is in the production of H_2SO_4.

Compounds.—Sulphur Dioxide, SO_2, an irrespirable, suffocating gas, is formed by S burning in the air, as in the lighting of a match. It is prepared for experiments by treating copper with strong H_2SO_4. On heating, SO_2 is evolved and may be collected like Cl by displacement. The action of the acid on the copper is represented thus:

$$Cu + 2H_2SO_4 = CuSO_4 + 2H_2O + SO_2.$$

SO_2 is more than twice as heavy as air. It is readily soluble in water, forming a solution which is believed to contain H_2SO_3, *sulphurous acid*. When

FIG. 48.

Bleaching by SO_2.

this is neutralized by bases it yields salts called *sulphites*. SO_2 extinguishes combustion. If our "chimney burns" at any time, we can easily quench the flame by burning a little S in the stove or on the hearth.

Uses.—SO_2 is used for bleaching silk, straw, and woolen fabrics, which are destroyed by Cl. Its action is very different from that of Cl, and depends upon the power it has of withdrawing O from substances or *reducing* them, as it is called. This takes place in the presence of moisture, and the result is that the SO_2 with H_2O and the O forms sulphuric acid, H_2SO_4. The reducing action of SO_2 is made use of in the manufacture of paper, to destroy the excess of Cl after bleaching:

$$SO_2 + Cl_2 + 2H_2O = H_2SO_4 + 2HCl.$$

SO_2 is called on this account an "anti-chlor." The bleaching effected by SO_2 is not permanent like that by Cl. The color is often restored by exposure to the air or by alkalies. S is also frequently employed to check fermentation, as when it is burned in a barrel before filling with new cider.

Sulphur Trioxide, SO_3, may be prepared by the oxidation of SO_2 in the presence of platinum sponge. It is often called *sulphuric anhydride.** If Nordhausen acid † be heated, the vapors may be condensed in a mass of silky, crystalline fibers of SO_3. This will show no acid reaction, will not redden blue litmus-paper, and, if the fingers are dry, can be molded like wax. If it be dropped into H_2O, it will hiss like a red-hot iron, and forming H_2SO_4, will exhibit all the properties of that corrosive substance.

Sulphuric Acid, *Oil of Vitriol,* is the king of the acids. It is of the utmost importance to the manufacturer and chemist, as it is used in the preparation of nearly all other acids, and many valuable compounds.

* An *anhydride* (without water) is a substance which, when dissolved in H_2O, will unite with its elements and form an acid.

† So named from the German town near which it was formerly made by the distillation of green vitriol (iron sulphate).

Preparation.—H_2SO_4 is such a strong acid that it can not be made by the action of some other acid on its salts, the sulphates, as HNO_3 is made from nitrates and HCl from chlorides. It is always made by oxidizing sulphurous acid, or what comes to the same thing, oxidizing SO_2 in the presence of moisture. If we burn a little S in a bottle, it will soon become filled with SO_2. Nitric acid, it will be remembered, easily parts with its O. So if we stir the SO_2 with a swab wet in aqua fortis, we shall quickly see the familiar NO_2 fumes, indicating that the acid has been decomposed and has given up O. Add a little water and shake the jar thoroughly. On testing the liquid with a few drops of a solution of barium chloride, a beautiful white precipitate will prove the presence of H_2SO_4.*

Fig. 40.

Making H_2SO_4.

The Manufacture of Sulphuric Acid on a large scale is based on the principle of the preceding illustration. The process is facilitated by the curious fact that nitric oxide (NO) has the property of acting as a carrier of O between the common air and H_2SO_3, whereby it can oxidize an almost indefinite quantity, thus forming H_2SO_4. S is burned in a current of air in furnaces A, A. In the stream of heated gas is suspended an iron pot, b, charged with a mixture of sodium nitrate and H_2SO_4. Vapors of HNO_3 are thus set free, and these pass on mixed

---

* The reaction in making the acid may be thus expressed :

$$2HNO_3 + 3SO_2 + 2H_2O = 3H_2SO_4 + 2NO.$$

The NO is at once converted into NO_2 by the O of the air in the bottle.

with SO_2 and excess of atmospheric air. The mingled gases pass into immense chambers, F, of sheet

FIG. 50.

Manufacture of H_2SO_4.

lead. A shallow layer of H_2O, d, covers the floor,
and the intermixture and chemical action of the
gases are further favored by the injection of jets of
steam, e, supplied from the boiler, G.

The chemical action which ensues may be explained as follows:—The nitric acid is quickly reduced
to nitric oxide, NO. This takes up an atom of O
from the air, becoming NO_2, and flies back to the
SO_2 which, with a molecule of H_2O becomes H_2SO_4,
a molecule of sulphuric acid. The NO once more
seeks the air and returns laden with O for the SO_2.
The weak sulphuric acid which collects on the floor
is drawn off and condensed by evaporation in lead
pans, and finally, when it begins to corrode the lead,
in platinum or glass stills. It is then put in large

FIG. 51.

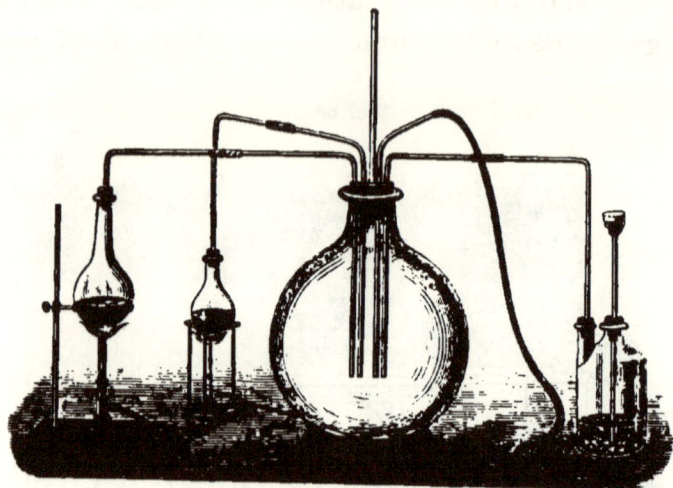

Making H_2SO_4.

bottles packed in boxes called *carboys*, when it is ready for transportation.

Properties. — It is a heavy, oily liquid, without odor, and when pure, colorless. The commercial acid is slightly colored by impurities. Its affinity for moisture is most remarkable. If exposed in an open bottle it gradually absorbs water from the air, and increases in bulk, sometimes even doubling its weight. It blackens wood and other organic substances, by taking away their H and O and leaving the C.* When mixed with H_2O, it produces much heat; 4 parts of acid to 1 of H_2O will boil a test-tube of water.† It commonly contains lead, which falls as a

---

* Strong oil of vitriol poured on a little loaf-sugar will convert it into black charcoal. Sugar consists of C, H, and O, and gives up the H and O to satisfy the acid.

† In mixing H_2SO_4 and H_2O, the H_2SO_4 should always be poured into the H_2O (and not the H_2O into the H_2SO_4) slowly and with constant stirring.

milky precipitate (PbSO₄) when the acid is diluted. It is the strongest of the acids, and will displace the others from their compounds. It stains cloth red, but the color can be restored by an alkali, if applied immediately. Its test is barium chloride, which forms a heavy white precipitate. In this way a drop of H_2SO_4 can be detected in a quart of H_2O.

FIG. 52.

Preparing H₂S.

Hydrogen Sulphide, H_2S, *Sulphuretted Hydrogen, Sulphydric Acid.*—This gas is produced in the decay of various organic substances, and is always found near cess-pools, drains, and sinks, turning lead paint black and emitting a disagreeable smell. It gives the characteristic odor to the mineral waters of Avon, Clifton, Sharon, and other celebrated sulphur springs. It is prepared by the action of dilute H_2SO_4 upon ferrous sulphide (FeS). The reaction is as follows:

$$FeS + H_2SO_4 = FeSO_4 + H_2S.$$

H_2S has the disgusting odor of rotten eggs. It burns with a pale blue flame, producing SO_2 and H_2O. It is very poisonous. The gas as well as its solution in H_2O are much used in the analytical laboratory to precipitate many of the metals as sulphides. Its test is lead acetate (sugar of lead), with which it forms black PbS (lead sulphide).

Carbon Disulphide, CS_2, is produced by passing the vapor of S over red-hot coals. It is a volatile, colorless liquid. The fact that a yellow, odorless solid thus unites with a black, odorless solid to form such a colorless, odoriferous liquid, illustrates very finely the transformations which may be effected by chemical affinity. CS_2 readily dissolves S, P, and I. It is largely used as a solvent for caoutchouc and as a means for recovering from wool the oil and fats with which they are treated. It is a powerful refractor of light, and is used for filling hollow, glass prisms employed in experiments with the solar spectrum. In its combustion it unites with O, forming CO_2 and SO_2.

PRACTICAL QUESTIONS.

1. If chlorine water stands in the sunlight for a time, it will only redden a litmus-solution. Why does it not bleach it?

2. Why do tinsmiths moisten with HCl, or sal-ammoniac, the surface of metals to be soldered?

3. How much HCl can be made from 25 kilos of common salt?

4. What weight of NaCl would be required to form 25 kilos of HCl.

5. HCl of a specific gravity of 1.2 contains about 40 per cent. of the gas. This is very strong commercial acid. What weight could be formed by the HCl gas produced in the reaction named in the preceding problem?

6. What is the difference between sublimation and distillation?

7. Why do eggs discolor silver spoons?

8. Explain the principle of hair-dyes.

9. Is it safe to mix oil of vitriol and water in a glass bottle?

10. What is the color of a sulphuric acid stain on cloth? How would you remove it?

11. What causes the milky look when oil of vitriol and water are mixed?

12. What is the chemical relation between animals and plants? Which perform the office of reduction, and which of oxidation?

13. How many pounds of S are contained in a cwt. of H_2SO_4?

14. How much O and H_2O are needed to change a ton of SO_2 to H_2SO_4?

15. How much O in a pound of H_2SO_4?

16. State the analogy between the compounds of O and S.

VALENCE.

In the Introduction it was stated that atoms had a certain property, in virtue of which each could hold a definite number of other atoms in combination, and that this property was called *valence*.

If we now review the formulas of some of the compounds with which we have become familiar, we shall be able to see more clearly just what is meant by this statement. We have had the following binary compounds of H :

$$HCl, \quad H_2O, \quad NH_3, \quad CH_4.$$

We see from these that the Cl, O, N, and C atoms differ from each other in respect to the number of H atoms which they hold in combination; O holds twice as many as Cl; N three times as many, and C four times as many as Cl or twice as many as O. No atom can hold a less number of any other atoms than Cl or H. That is to say, they have the property of valence in the lowest degree. On this account Cl,

H, and other atoms which can hold but one of them are called *univalent;* atoms which, like O, can hold two of H or two unit atoms in combination, are called *bivalent;* atoms like N, which can hold three unit atoms, are called *trivalent;* and atoms like C, which hold four, are called *quadrivalent.* Thus, F is seen to be univalent (HF); S, bivalent in H_2S and quadrivalent in SO_2 (since each O atom is equivalent to two atoms of H); P trivalent in PH_3.

When the H of acids is replaced by metals in the formation of salts, a univalent metal may take the place of each atom of H, as in NaCl, $NaNO_3$, Na_2SO_4; or a bivalent metal may take the place of two atoms of H, as in $MnCl_2$, $Cu2NO_3$, $PbSO_4$; and so on.

Acids which contain but one atom of replaceable H are called *monobasic* acids. Such acids can form only one kind of salt with the same metal. Acids which have two (*bibasic*) or more (*tribasic*, etc.) replaceable atoms of H can form two or more classes of salts, in some of which only a part of the H is replaced by a metal; in others, all. Thus, H_2SO_4, a bibasic acid, forms Na_2SO_4, and $NaHSO_4$. The former is called *normal* sodium sulphate, the latter *acid* sodium sulphate, because it still contains replaceable H.

PHOSPHORUS.

Symbol, P.....Atomic Weight, 31.....Specific Gravity, 1.83.

THE name Phosphorus signifies *light-bearer*, given because this substance glows in the 'dark. It was called by the old alchemists "the son of Satan."*

Occurrence.—It exists in combination with O and metals in a number of minerals and rocks, and by their decay passes into the soil, is taken up by plants, is then stored in their seeds (wheat, corn, oats, etc.), and finally passes into our bodies. As calcium phosphate ($Ca_3 2PO_4$, phosphate of lime), it is a prominent constituent of our bones.† Phosphorus is so necessary to the operation of the brain that a saying has become current, "Without phosphorus, no thought."‡

FIG. 53.

Manufacture of Phosphorus.

Preparation.—It is prepared in considerable quanti-

---

* The following singular event is said to have occurred many years before the reputed discovery of phosphorus by Brandt in 1669. A certain Prince San Severo, at Naples, exposed some human skulls to the action of several re-agents, and then to the heat of a furnace. From the product he obtained a substance which burned for months without apparent loss of weight. San Severo refused to divulge the process, as he wished his family vault to be the only one to possess a "*perpetual lamp*," the secret of which he considered himself to have discovered.

† "Of phosphorus every adult person carries enough (1¾ lbs.) about with him in his body to make at least 4,000 of the ordinary two-cent packages of friction matches, but he does not have quite sulphur enough to complete that quantity of the little incendiary combustibles."—NICHOLS' *Fireside Science.*

‡ See an article by Prof. Atwater in *The Century*, for June, 1887, page 249.

ties from bones. These are first calcined to whiteness to burn out the animal matter, then treated
with H_2SO_4, which changes the calcium phosphate
into a compound which is reduced at a high temperature by C. The phosphorus which distils as a
vapor is condensed under H_2O.

Properties.—It is a waxy, translucent solid, at all
temperatures above 0° C. emits a feeble light, melts
at 44° C., and ignites at a little higher temperature.
It should be handled with the utmost care, always
kept and cut under H_2O, never used except in very
small quantities, and never held in the hand. Its
burns are deep and dangerous. It is poisonous, and
its vapor produces horrible ulcerations of the jaw-
bone in workmen who use it.

Red Phosphorus.—Heated for several hours at a
temperature of about 250° C., in a close vessel, the
melted phosphorus changes into a brick-red solid,
and loses all its former properties. It is now insoluble in CS_2, which can be used to dissolve out every
trace of the common form. Its specific gravity is
increased to 2.14. It can be handled with impunity,
carried in the pocket like so much snuff, and even
heated to nearly 260° C. without taking fire. At a
little over 260° C., however, it changes into the common form and bursts into a blaze.

Uses.—Matches.—The principal use of phosphorus
is in the manufacture of matches. 1. *The Lucifer
Match.*—The bits of wood are first dipped in melted
S and dried; then in a paste of phosphorus, niter,
and glue, which completes the process. The object

of the niter is to furnish O to quicken the combustion. Instead of this, potassium chlorate is sometimes used; it can be recognized by a crackling sound and jets of flame when ignited. The tips are colored by red-lead, or Prussian blue, mixed in the paste. When a match is burned, the reaction is as follows: first, the phosphorus ignited by friction burns, forming P_2O_5;* this produces heat enough to inflame the S, which makes SO_2; lastly, the wood takes fire, and forms CO_2 and H_2O. Thus there are four compounds produced in the burning of a single match.

2. *The Safety Match.*—The pieces of wood are dipped into melted paraffine (see p. 191) and dried. They are then capped with a paste of potassium chlorate, sulphide of antimony, powdered glass, and gum-water. They ignite only when rubbed on a surface covered with a mixture of red phosphorus and powdered glass.

Phosphorescence is the property of emitting light, without the high temperature which accompanies ordinary burning. We have seen that P glows in the dark, but it is by no means the only substance which presents this phenomenon. The luminous appearance of putrefying fish and decayed wood is well known. The latter is sometimes called "fox-

---

* The burning phosphorus produces a very luminous flame, because of the reflection of light from the dense vapor (P_2O_4). The following experiment is very suggestive in this connection: Ignite a bit of phosphorus placed upon a sheet of white paper. The paper will be blackened just where the phosphorus lay, but will not take fire; and after the flame is extinguished, one can write upon it with pen and ink, close to the edge of the charred portion.

fire." The "glow-worm's fitful light" is associated
with our memory of beautiful summer evenings. In
the West Indies, fire-flies are found that emit a green
light when resting, and a red one when flying. They
are so brilliant that one will furnish light enough

Fig. 54.

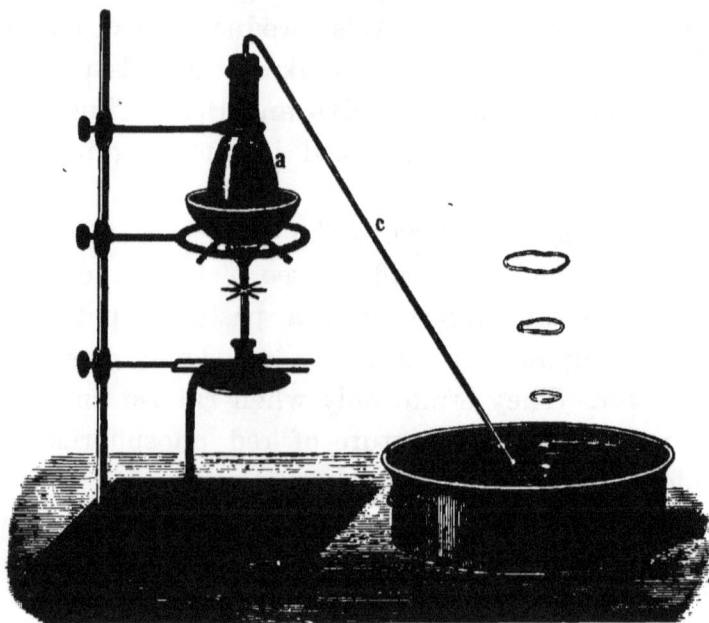

Preparation of PH₃.

for reading. The natives wear them for ornaments
on their bonnets, and illuminate their houses by
suspending them as lamps. The ocean occasionally
takes on strange colors, and the sailor finds his vessel
plowing at one time apparently a furrow of fire, and
at another one of liquid gold. The water is all
aglow, and the flames seem to leap and dance with

the waves or the motion of the ship. The phenomenon is produced by multitudes of animalcules which frequent certain seas.*

Compounds. — Hydrogen Phosphide, PH₃, *Phosphuretted Hydrogen*, is a poisonous gas, remarkable for its disgusting odor, for igniting spontaneously on coming to the air, and for the singular beauty of the rings formed by its smoke. It is prepared by heating in a retort a strong solution of potash containing a few bits of phosphorus. It has been thought by some that the Will-o'-the-wisp, Jack-o'-the-lantern, etc., as seen near grave-yards and in swampy places, are produced by this gas coming off from decaying substances, and igniting as it reaches the air.

ARSENIC.

Symbol, As.... Atomic Weight, 75... Specific Gravity, 5.7.

Volatilizes without fusion at about 450° C.

As is a brittle, steel-gray metal,† commonly sold, when impure, as *cobalt*.‡ If heated in the open air

* Many substances, after having been exposed to the light, will shine for some time when removed into the darkness. Thus, a dial coated with the so-called *Luminous Paint* (calcium sulphide) will show the time at night.

† Arsenic very much resembles phosphorus in its general properties, and is therefore classified with it, but it conducts electricity moderately, and has a high brilliancy. It is intermediate between the non-metals and the metals.

‡ Cobalt is a reddish-white metal, found in combination with arsenic. Co received its name from the miners, because its ore looked so bright that they thought they would obtain something valuable; but when, by roast-

it gives off the odor of garlic, which is a test of As.

Arsenic Trioxide, As_2O_3.—This is the well-known "*ratsbane*," and is sometimes sold as simply "arsenic," or "white arsenic."

Preparation.—It is made in the Harz and elsewhere, by roasting arsenical ores at the bottom of a tower, above which is a series of rooms through which the vapors ascend, and pass out at a chimney in the top. The As burns, forming As_2O_3, which collects as a white powder on the walls and floors of the chambers above.*

Properties.—"Arsenic" is slightly soluble in H_2O, and has a weak metallic, faintly sweetish taste. It is a powerful poison, doses of two or three grains being fatal, although an over-dose acts as an emetic. It is an antiseptic, and so in cases of poisoning frequently attracts attention by the preservation of parts of the body, even twenty or thirty years after

ing, it crumbled to ashes, they believed themselves mocked by the evil spirit (Kobolt) of the mines. The oxide of cobalt makes a beautiful blue glass, which, when ground fine, is called *smalt*. It is used for tinting paper, and by laundry women to give the finished look to cambrics, linen, etc. Its impure oxide, called *zaffer*, imparts the blue color to common earthenware and porcelain. The chloride ($CoCl_2$) is used as a sympathetic ink. Letters written with a dilute solution of it are invisible when moist with the H_2O absorbed from the air, but on being dried at the stove, again become blue. If the paper be laid aside the writing will disappear, but may be revived in the same manner. A winter landscape may be drawn with India-ink, the leaves being added with this ink. On being brought to the fire it will bloom into the foliage of summer.

* Its removal is a work of great danger. The workmen are entirely enveloped in a leathern dress and a mask with glass eyes; they breathe through a moistened sponge, thus filtering the air of the fine particles of arsenic floating through it. Yet, in spite of all these precautions, they rarely live beyond forty.

the murder has been committed. The antidote is milk or white of egg.*

Marsh's Test.—There is no other poison which is so easily detected. Prepare a flask for the evolution of H. Ignite the jet of gas, and hold in the flame a

FIG. 55.

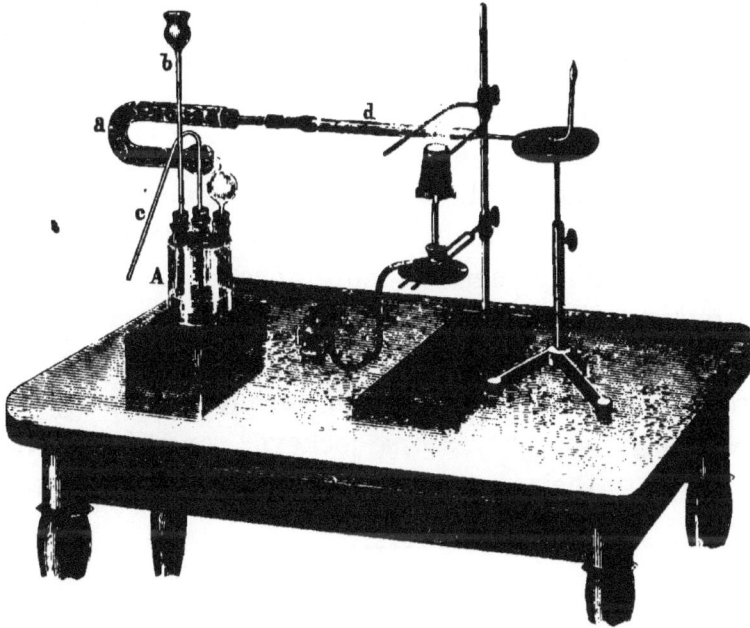

Marsh's Test.

cold porcelain dish. If it remains untarnished, the materials contain no As. Now pour in through the funnel-tube a few drops of a solution of As;† the color of the flame will be seen to change almost

---

* The exact chemical antidote is hydrated ferric oxide. In this, as in most other cases of poisoning, where the antidote is not at hand, an emetic should be taken at once—a tea-spoonful of mustard in a glass of warm water, or even a quantity of soap-suds. (See " Physiology," page 209.)

† This is made by dissolving a little As_2O_3 in HCl.

instantly, and a copious deposit of As will be formed on the dish. If the tube through which the gas is passing be heated (see fig. 55), a metallic mirror of arsenic will appear just beyond the heated place.* The gas formed in this experiment—arseniuretted hydrogen—is very poisonous indeed, and the utmost care should be used to prevent its inhalation.

Arsenic-Eating.—It is said that the peasants in a portion of Styria are accustomed to eat As, both fasting and as a seasoning to their food. A very minute portion will warm, stimulate, and aid in climbing lofty mountains. The arsenic-eaters are described as plump and rosy, and it is said that the young people resort to this dangerous substance, as a species of cosmetic. They begin with small doses, which are gradually increased; but if the person ceases the practice, all the symptoms of arsenic poisoning immediately appear. Horse-jockeys sometimes feed arsenic to their horses to improve their flesh and speed.

BORON.

Symbol, B........Atomic Weight, 11.

Boron is found in nature in combination with H and O as boric acid, and as borates. Boric acid is

---

* In a case of poisoning the contents of the stomach would, of course, be substituted for the solution of As, the organic matter first being destroyed, and other tests besides these would be employed. We can imagine with what care a chemist would conduct the examination, and with what intense anxiety he would watch the porcelain dish as the flame played upon it, hesitating, and dreading the issue, as he felt the life of a fellow-being trembling on the result of his experiment.

abundant in the volcanic districts of Tuscany.* Along the sides of the mountains, series of basins are excavated and filled with cold water from the neighboring springs. Into these basins the jets of steam, charged with boric acid, are conducted. The H_2O absorbs the acid, and itself becomes heated to the boiling-point. It is then drawn off into the next

Fig. 56.

Preparing Boric Acid in Thibet.

lower basin. This process is continued until the bottom one is reached, when the solution runs into leaden pans heated by the steam from the earth; here the H_2O is evaporated, and the boric acid collected.

Borax ($Na_2B_4O_7$, $10H_2O$) is a salt of this acid. It is a natural production, formerly obtained by the

* Throughout an area of nearly thirty miles, is a wild, mountainous region, of terrible violence and confusion. The surface is ragged and blasted. Every-where there issue from the ground jets of steam, filling the air with offensive odors. The earth itself shakes beneath the feet, and frequently yields to the tread, ingulfing man and beast. "The waters below are heard boiling with strange noises, and are seen breaking out upon the surface. Of old, it was regarded as the entrance to hell. The peasants pass by in terror, counting their beads and imploring the protection of the Virgin."

drying of certain lakes in Thibet, but since found abundantly in California and Nevada. When dissolved in alcohol, boric acid gives to the flame a peculiar green tint. This is an easy test of the presence of this acid. In order to make this test with borax, H_2SO_4 must be added to set the boric acid free, for the salts of boric acid do not color the flame. The salt is employed in welding. It dissolves the oxide of the metal, and keeps the surface bright for soldering. It has mild cleansing properties due to its power of dissolving oils and resinous substances and is used in washing.*

SILICON.

Symbol, Si.... Atomic Weight, 28.... Specific Gravity, 2.49.

Occurrence.—Silicon is found in combination with O as silica (silicic anhydride, SiO_2), commonly called silex or quartz. So abundant is this oxide that it probably comprises nearly one half of the earth's crust. (See "Geology," p. 40.) It forms beautiful crystals and some of the most precious gems. When pure, it is transparent and colorless, as in rock

---

* Borax is extensively used in "blow-pipe analysis." When melted with chromium oxide, it gives an emerald green; with cobalt oxide, a deep blue; with copper oxide, a pale green; and with manganese oxide, a violet.—The Salt, or Alkali, Marshes of Nevada, contain extensive deposits of sodium chloride, sodium carbonate, sodium borate, calcium borate, etc. There are hundreds of acres covered to a depth of nearly two feet with crude, semi-crystalline borax. See an article entitled "Borax in America," in the *Popular Science Monthly*, July, 1882.

crystal. Jasper, amethyst, agate, chalcedony, blood-stone, chrysoprase, sardonyx, etc., are all common flint-stone or quartz, colored with some metallic oxide. The opal is only SiO_2 and H_2O. Sand is mainly fine quartz, which, when hardened and ce-mented, we call sandstone. Yellow or red sand is colored by iron-rust.

Properties of SiO_2.—It is tasteless, odorless, and colorless. It seems very strange to call such an inert substance an anhydride; yet it unites with the al-kalies, neutralizes their properties, and forms a large class of salts known as the silicates, which are found in the most common rocks.—*Example:* feldspar, found in granite.

Silica in Soil and Plants.—Silica is insoluble in H_2O, unless it contains some alkali. When the sili-cates, so abundant in rocks, disintegrate and form soil, the alkali and silica are both dissolved in the water, and taken up by the roots of plants. We see the silex on the surface of scouring-rushes and sword-grass, which cut the fingers if handled care-lessly. It gives stiffness to the stalks of wheat and other grains, and produces the hard, shiny surface of bamboo, corn, etc.

Petrifaction.—Certain springs contain large quan-tities of some alkaline carbonate; their waters, there-fore, dissolve silica abundantly. If we place a bit of wood in them, as fast as it decays, particles of silica will take its place and thus petrify the wood. The wood has not been *changed to* stone, but has been *replaced by* stone.

Compounds.—The Silicates.—Glass* is a mixture of several silicates. There are four varieties used in the arts. 1. *Window or plate glass* is composed of silicates of calcium and sodium. It is made by heating white sand, sal-soda, and lime in clay crucibles for about forty-eight hours, when the materials fuse and combine into a double silicate. The Ca hardens and gives luster; the Na renders the glass fusible, but imparts a greenish tint. 2. *Bohemian glass* consists of silicates of calcium and potassium. It can withstand a very high temperature without softening, and is hence of great importance to the chemist. 3. *Flint-glass†* or *crystal* contains silicates of potassium and lead. The latter is used in large quantities and produces a soft, lustrous glass, which can be ground into imitation gems, tableware, chandelier pendants, prisms, etc. 4. *Green bottle-glass* is made of silicates of calcium, sodium, aluminium, and iron. The last gives the opaque green of the common junk bottle.

* Glass was known to the ancients. Hieroglyphics, that are older than the sojourn of the Israelites in Egypt, represent glass-blowers at work, much after the fashion of the present. In the ruins of Nineveh, articles of glass, such as vases, bowls, etc., have been discovered. Mummies, three thousand years old, are adorned with glass beads. The inventor is not known. Pliny tells us that some merchants, once encamping on the seashore, found in the remains of their fire bits of glass, formed from the sand and ashes of the sea-weed by the heat; but this is impossible, as an open fire is not sufficient to melt these materials. In the fourth century B.C., the glass-works at Alexandria produced exquisite ornaments, with raised figures beautifully cut and gilded. But in the twelfth century A.D., glass was still so costly in England that glass windows were thought very magnificent; and, as late even as about 1500, when the great Earl of Northumberland left one of his houses for a time, he was careful to have the glass of the windows taken down and packed for safe-keeping.

† So called because pulverized flint was formerly used for sand.

Coloring Glass.—A small quantity of some metallic oxide melted with the glass furnishes any tint desired: Co gives a beautiful sapphire blue; Au or Cu, a ruby-red; Mn, a violet; U, a yellow; As, a soft white enamel, as in lamp-shades; and Sn, a hard enamel, as in watch-faces.

Annealing Glass.—If the glass utensils were cooled immediately, they would be found extremely brittle, and would drop to pieces in the most unaccountable way. The heat of the hand or a draft of cool air would sometimes crack off the thick bottom of a tumbler. They are therefore cooled very gradually for days, which allows the particles to assume their natural place, and the molecular attractions to become equalized.*

Ornamental Ware.—Venetian balls or paper weights are made by arranging bits of colored glass in the form of fruits, flowers, etc., and then, inserting them in a hollow globe of transparent glass, still hot, the workman draws in his breath, and the pressure of the air above collapses the globe upon the colored glass, and leaves a concave surface in the opposite side of the weight. The lens form always magnifies the size of the figures within.

Tubes and , Beads.—In making glass tubing, the workman inserts his iron blowing-tube in a pot of melted glass, and gathers upon the end a suitable amount; drawing this out, he blows into the tube, swelling the glass into a globular form. Another

* This principle is beautifully illustrated by the toy known as the "Prince Rupert's Drop." (See "Physics," p. 46.)

dip into the pot and another blow increase its size, until at last a second workman attaches an iron rod to the other end. The two men then separate at a rapid pace. The soft glass globe diminishes in size as it lengthens, until at last it hangs between them a glass tube of a hundred feet in length, and perhaps only a quarter of an inch in diameter.

For making beads, glass tubes are cut in short pieces, and then worked about in a mixture of wet ashes and sand, until they are filled. They are next put with loose sand in a cylinder rapidly revolving over a hot furnace. The heat softens the glass, but the mixture within presses out the sides, and the sand grinds the edges, until at last the beads become round and perfect, and are taken out ready for market.

THE METALS.

THE METALS OF THE ALKALIES.

K, Na, Li, Rb, Cs.

POTASSIUM,

Symbol, K....Atomic Weight, 39....Specific Gravity, 0.87.

Source.—K occurs widely distributed in various rocks, which by their decomposition furnish it to the plants.* When these are burned it remains as potassium carbonate in the ashes. It also is found in considerable deposits in the form of chloride and nitrate.

Preparation.—This metal was discovered by Sir Humphry Davy, in 1807. On passing the current of a powerful electric battery through potash, the globules of the K appeared at the negative pole. The metals Na, Ba, Sr, and Ca, were afterward separated in the same manner. This discovery constituted a most important epoch in chemistry. K is now

---

* " An acre of wheat, producing twenty-five bushels of grain and 3,000 lbs. of straw, removes about 40 lbs. of potash in the crop. An acre of corn, producing 100 bushels, removes in kernel and stalk 150 lbs. of potash and 80 lbs. of phosphoric acid. An acre of potatoes, yielding 300 bushels, will remove in tubers and tops 400 lbs. of potash and 150 lbs of phosphoric acid. A pound of wheat holds a quarter of an ounce of mineral substances, and a pound of potatoes one eighth of an ounce."—NICHOLS.

prepared by decomposing potassium carbonate by means of charcoal in iron bottles, at an intense heat. The green vapors of K distil and are condensed in receivers of naphtha, and CO passes off as a gas, $K_2CO_3 + 2C = 2K + 3CO$. It is a difficult and dangerous process. The vapor takes fire instantly on contact with air or water. It also absorbs CO, and the compound, if kept, becomes powerfully explosive. To prevent this danger, the K is immediately redistilled.

Properties.—K is a silvery-white metal, soft enough to be spread with a knife, and light enough to float like cork. Its affinity for O is so great that it is always kept under the surface of naphtha, which contains no O. K, when thrown on H_2O, decomposes it with great energy (see page 38).

Compounds.—Potassium oxide, K_2O, has so great an affinity for H_2O that the anhydrous form is rarely prepared. Its hydrate, potassium hydroxide, KOH,* is a white solid made from potassium carbonate by the action of slaked lime. It is the most powerful alkali. It neutralizes the acids, and turns red litmus to blue. It is used to cauterize the flesh, and is hence commonly called " caustic potash." It dissolves the cuticle of the finger which touches it, and so has an unctuous feel. It unites with grease, forming soft-soap, in the manufacture of which it is extensively used.

Potassium Carbonate, K_2CO_3, *Pearlash,* " *Carbonate of Potash,*" is obtained in the following manner: Potash exists in plants, combined with various acids, such as tartaric, malic, oxalic, etc. When the wood

---

* $K_2O + H_2O = 2KOH$, or 2 molecules of potassium hydroxide.

is burned, the organic salts are decomposed by the heat, and K remains chiefly as K_2CO_3. The ashes are then leached and the *lye* is evaporated, when the K_2CO_3 crystallizes. This forms the "potash" of commerce. When refined, it is called "pearlash." Where wood is abundant, immense quantities are burned solely for this product. Birch gives the purest potash, while the leaves of a tree furnish twenty-five times as much as the heart.*

Large quantities of potash are also obtained from potassium chloride and sulphate, by a process which is exactly like that described on p. 135 for making soda.

Acid Potassium Carbonate, † $HKCO_3$, *Saleratus,* "*Bicarbonate of Potash,*" is prepared by passing CO_2 through a strong solution of potassium carbonate.

Potassium Nitrate, KNO_3, *Nitrate of Potash, Saltpeter, Niter.*—This salt is found as an efflorescence on the soil in tropical regions, especially in India. It is obtained thence by leaching.‡ It is formed arti-

---

* Vast deposits of potassium salts have been opened up to us at the Stassfurth salt mines in Germany, the supply from which is more than from the wood-ash sources of the whole world. "Only about 13,000 tons of potash were sent to market from the United States and British America in 1870, and yet from Stassfurth, where a dozen years ago it was not supposed that a single ton could be produced, 30,000 tons of potassium chloride were manufactured and supplied to consumers upon both continents during the following year. • The surface salts at these mines, which hold the potash, are practically inexhaustible, and millions of tons will be supplied in succeeding years."—*Fireside Science.*

† The molecule of carbonic acid is H_2CO_3. In potassium carbonate, K_2CO_3, both the atoms of H contained in the carbonic acid are replaced by the metal K ; in hydrogen potassium carbonate, $HKCO_3$, only one atom of H is thus replaced.

‡ It was manufactured in the Mammoth Cave, Kentucky, during the war of 1812. The remains of the works, and even the deep ruts of the wagon-wheels, are still to be seen, preserved in the pure, still air.

ficially by piling up great heaps of mortar, refuse of sinks, stables, etc. "In about three years, these are washed, and each cubic foot of the mixture will furnish four or five ounces of saltpeter." It dissolves in about three and a half times its weight of cold H_2O.

Properties and Uses.—It is cooling and antiseptic; hence it is used with common salt (NaCl) for preserving meat. It parts readily with its O, of which it contains nearly 48 per cent., and deflagrates with charcoal brilliantly. Every government keeps a large supply on hand for making gunpowder, in the event of war. Gunpowder is now composed of about six parts, by weight, of niter, and one each of charcoal and sulphur—the proportion varying with the purpose for and the country in which it is made. Its explosive force is due to the expansive power of the gases formed. At the touch of a spark the saltpeter gives up its O to burn the S and C. The reaction that ensues may be approximately represented as follows: $2KNO_3 + S + 3C = K_2S + N_2 + 3CO_2$.

Besides N and CO_2, smaller quantities of other gases are formed, which, with the sudden increase of temperature (to about 2,200°C.), expand till they occupy at least 1,500 times the space of the powder. The bad odor of burnt powder is due to the slow formation of H_2S in the residuum. Fire-works are composed of gunpowder ground with additional C and S, and some coloring matter. Zinc filings produce green stars; steel filings, variegated ones.

$Sr2NO_3$ tinges flame with crimson; $Ba2NO_3$ with green. Salts of copper give a blue, and camphor·a pure white flame.

Potassium Chlorate, $KClO_3$, is a white, crystallized salt much used in preparing oxygen, making matches, fire-works, etc. It is a powerful oxidizing agent.*

Potassium Bichromate, † $K_2Cr_2O_7$, is a red salt highly valued in dyeing, calico-printing, and photo-lithography. If we mix a solution of this salt and one of sugar of lead, a yellow-colored precipitate will be formed, known in the arts as chrome-yellow (lead chromate). $K_2Cr_2O_7$ is a strong oxidizing agent, being readily reduced in an acid solution to a salt of Cr.

----·•·----

SODIUM.

Symbol, Na.... Atomic Weight, 23.... Specific Gravity, 0.972.

THIS metal is found principally in common salt. Its preparation is similar to that of K, but is more

---

* *Examples:* 1. Put in a porcelain crucible as much $KClO_3$ as will lie on the point of a knife-blade, and half as much S. On grinding with the pestle, rapid detonations will ensue. 2. Place in a wine-glass five or six pieces of phosphorus as large us a grain of wheat, and cover with crystals of $KClO_3$. Fill the glass two thirds full of H_2O. By means of a pipette, or a glass funnel, introduce into immediate contact with the $KClO_3$ a few drops of strong H_2SO_4. A violent chemical action will immediately occur, and the phosphorus will burn under the water with vivid flashes of light.

† Chromic anhydride (Cr_2O_3) is an oxide of chromium (*chroma*, color), a metal prized only for its numerous brilliantly colored compounds. It is rather rare, and mainly found in chrome iron-stone ($FeO.Cr_2O_3$).

easily managed. It is very like K in appearance, properties, and reaction. It decomposes water energetically, but does not take fire like the K, unless the water is warm. The test of all the soda salts is the yellow tint which they give to flame.

Compounds. — Sodium Chloride, NaCl, *Common Salt*, is a mineral substance absolutely necessary to the life of human beings and the higher orders of animals. It does not enter into the composition of tissue, but is essential to the proper digestion of the food, and to the removal of worn-out matter. (See "Physiology," p. 137.) Among the many cruel punishments inflicted in China, deprivation of salt is said to be one, causing at first a most indescribable longing and anxiety, and finally a painful death. As salt is so universally necessary, it is found every-where. Our Father in fitting up a home for us, did not forget to provide for all our wants. The quantity of salt in the ocean is said to be equal to five times the mass of the Alps. Salt lakes are scattered here and there; saline springs abound; and besides these, in the earth are stored great mines, probably produced by the evaporation of salt lakes in some ancient period of the earth's history. Near Cracow, Poland, is a bed five hundred miles long, twenty miles wide, and a quarter of a mile thick. In Spain, and lately in Idaho, it has been quarried out in perfect cubes, transparent as glass, so that a person can read through a large block.

Preparation.—On the sea-shore it is manufactured by the evaporation of sea-water, each gallon contain-

ing about four ounces.* At Syracuse, New York,
near by and underneath the Onondaga Lake, is apparently a great basin of salt-water, separated from
the fresh-water above by an impervious bed of clay.
Upon boring through this, the saline water is
pumped up in immense quantities. The H_2O is
evaporated by heating in large iron kettles over a
fire, or in shallow, wooden vats by exposure to the

FIG. 57.

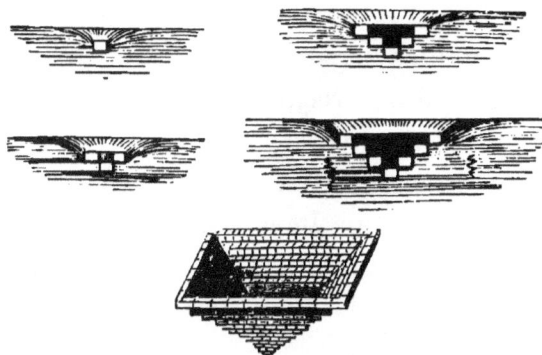

Hopper form of salt crystals.

sun—whence the name "solar salt." If boiled down
rapidly, fine table-salt is made; if more slowly,
coarse salt, as large crystals have time to form.
Frequently they assume a "hopper shape," one cube
appearing, then others collecting at its edges, and
gradually settling, until a hollow pyramid of salt-
cubes, with its apex downward, is formed.

* "Salt is soluble in less than three times its weight of H_2O. It is
scarcely more soluble in hot than cold H_2O, and a *saturated* solution (one
containing all it will dissolve) has about 36 per cent. Sea-water contains
about 3 per cent. Sodium carbonate was formerly obtained from the ashes
of sea-plants, as potassium carbonate is now from the ashes of land-plants."
—ROSCOE.

Uses.—NaCl is used largely as a fertilizer, for preserving meats and fish, and for preparing Cl, HCl, and the various compounds of Na.

Sodium Hydroxide, Caustic Soda, NaOH, is prepared from sodium carbonate by the action of milk of lime :

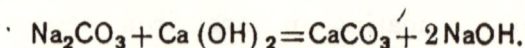

$$Na_2CO_3 + Ca\,(OH)_2 = CaCO_3' + 2\,NaOH.$$

It resembles KOH, but is less powerful in its chemical action. It is largely used in soap-making, and other technical operations.

Sodium Sulphate (Na_2SO_4, $10H_2O$), *Glauber's Salt,* named from its discoverer, is made in great quantities from NaCl, as the first stage in the manufacture of sodium carbonate. It is remarkably efflorescent, the salt, by exposure to the air, losing its ten molecules of H_2O.* It has a bitter, saline taste, and is used in medicine.

Sodium Carbonate (Na_2CO_3, $10H_2O$), *Sal-soda,* is used extensively in the arts. It is, therefore, of great importance to all consumers of soap, glass, etc., that it should be manufactured as cheaply as possible. Leblanc's process of making it from NaCl is the one generally adopted. The operation comprises two

---

* *Experiment:* Make a saturated solution of sodium sulphate in warm water, and with it fill a bottle. Stuff cotton loosely into the neck of the bottle, and let it stand. The salt will remain for months without crystallizing; but if it be taken up, and shaken ever so little, the whole mass will instantly form into crystals, so filling the bottle that not a drop of water will escape, even if it be inverted. Should there be any hesitation in crystallizing at the moment, drop into the bottle a minute crystal of the salt, and the effect will instantly be seen in the darting of new crystals in every direction.

stages: Changing, 1. NaCl into Na_2SO_4; and, 2. Na_2SO_4 into Na_2CO_3.

1. A mixture of NaCl and H_2SO_4 is heated. Na_2SO_4 is formed with a copious evolution of HCl. The fumes of this gas are conducted into the bottom of a vertical flue filled with pieces of coke wet with constantly falling H_2O. The gas is here absorbed, and a weak muriatic acid formed in great quantities.*

2. The Na_2SO_4 is heated with chalk ($CaCO_3$) and char-coal. The C deoxidizes the Na_2SO_4, changing it into Na_2S. The metals of the Na_2S and the $CaCO_3$ change places, forming Na_2CO_3 and CaS. Out of this mass, called from its color "black-ash," the Na_2CO_3 is dis-solved,† and then evaporated to dryness, making the "soda-ash" of commerce.

Hydrogen Sodium Carbonate ($HNaCO_3$), "*Bicar-bonate of Soda*," is the "soda" of the cook-room. It is prepared by the action of CO_2 on sodium car-bonate. The CO_2 may be easily liberated by the action of an acid. (See p. 244.)

Sodium Nitrate, $NaNO_3$, occurs in large deposits as *Chili Saltpeter*. It attracts moisture from the air, and so can not take the place of the more expensive KNO_3 in the manufacture of gunpowder.

* This acid was formerly allowed to escape, causing the destruction of all vegetation in the neighborhood. It is now, however, absorbed so per-fectly that the gases which escape from the top of the chimney will not render turbid a solution of silver nitrate (see page 97), showing that there is not a trace of the acid left.

† The insoluble residuum of CaS, and the superfluous coal, form around the alkali works a mountain of waste. Attempts have been made to ex-tract the S, and at the Paris Exposition large blocks thus obtained were exhibited.

AMMONIUM.

Symbol, NH_4.

THIS is a compound which has never been separated, but it is generally thought to be the base of the salts formed by the action of the acids upon the alkali ammonia, which closely resemble the corresponding salts of K. The analogy between it and the simple metals is so very striking* that it is considered a compound metal, acting the part of a simple one, as Cy does that of a compound halogen (see p. 74).†

Compounds.—Ammonium Chloride, NH_4Cl, *Sal-ammoniac*, is prepared from the ammoniacal liquor of the gas-works. (See p. 73.) It is obtained either as a fine crystalline meal or as a tough fibrous, crystalline mass. In neither form does it reveal any trace of the pungent ammonia, yet this can easily be set free, as we have already seen (p. 35). Sal-ammoniac is soluble in H_2O. It is used in medicine, in the preparation of NH_3 and the salts of NH_4, in dyeing,

---

* When NH_3 is dissolved in H_2O, forming NH_3HOH the compound may be represented as (NH_4) OH. Comparing this with the formula for caustic potash, KOH, we see that the group of elements NH_4 corresponds to the K. Thus we may call a solution of NH_4, ammonium hydroxide or "caustic ammonia," as one of potash is a potassium hydroxide or "caustic potash." Both act as powerful bases, neutralize the acids and form soaps.

† The following experiment is thought by some to be an additional proof of the metallic nature of this compound substance. Heat moderately in a test-tube half a fluid-dram of Hg with a piece of Na the size of a pea. The two metals will combine, forming a pasty amalgam. When cold, pour over it a solution of sal-ammoniac. The amalgam will immediately swell up to eight or ten times its original bulk, *retaining, however, its metallic luster.* The ammonium can not be separated from the amalgam, since, on heating, it decomposes, and on being thrown into water, H is set free and NH_3 formed.

and also in soldering, as it dissolves the coating of the oxide of the metal and preserves the surfaces clear for the action of the solder. It is also much used now in certain electric batteries.

Ammonium Carbonate, *Sal-volatile, Smelling Salts,* is prepared by the action of chalk upon sal-ammoniac. It is a sesquicarbonate, but by the constant loss of NH_3 it becomes crusted with a spongy coat of the "bicarbonate," hydrogen ammonium carbonate $(NH_4)HCO_3$. It is largely used by bakers in raising cake. (See p. 246).

Ammonium Nitrate (NH_4NO_3) may be readily formed by cautiously adding dilute HNO_3 to aqua ammonia until the liquid becomes neutral, and then evaporating. Long, needle-shaped crystals will form. Thus two fiery liquids combine to produce a solid having no resemblance to either of them. By heat this salt may be converted into H_2O and N_2O. (See p. 32.) All the salts of ammonium are volatile and decomposed by heat.

THE RARE METALS OF THE ALKALIES.

LITHIUM (Li), Rubidium (Rb), and Cæsium (Cs) are much rarer than K and Na. Li is the lightest of all metals (specific gravity 0.59) and it has the lowest atomic weight (7). The metal and its salts resemble Na and its salts, while Rb and Cs are more like K. Cs was the first metal discovered by spectrum analysis. (See page 147.)

METALS OF THE ALKALINE EARTHS.

Ca, Ba, and Sr.

CALCIUM.

Symbol, Ca.... Atomic Weight, 40.... Specific Gravity, 1.57.

Ca exists abundantly in limestone, gypsum, and in the bones of the body.* It commonly occurs, in nature, as sulphate or carbonate; and, in commerce, as oxide.

Compounds.—Calcium Oxide (CaO), *Caustic* or *Quicklime*, is obtained by heating limestone ($CaCO_3$) in large kilns. The CO_2 is driven off by the heat, and the CaO is left as a white solid.

Fig. 58.

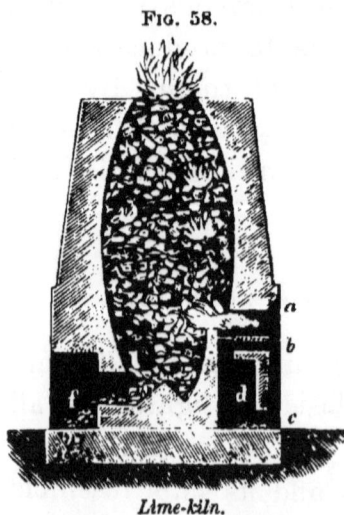

Lime-kiln.

Fig. 58 shows a form of lime-kiln in which the process is continuous. At *a*, *b*, *c*, are the doors for the fuel, ash-pit, etc. The kiln is fed at the top with limestone from time to time, while the lime, settling at the bottom, is taken out at *f* as fast as it is formed.

---

* "There are 5 lbs. of phosphate of lime, one of carbonate of lime, and 3 oz. of fluoride of calcium in the body of an adult weighing 154 lbs."— NICHOLS.

Properties.—CaO has such an affinity for H_2O, that fifty-six pounds of lime will absorb eighteen pounds of H_2O, forming $Ca(OH)_2$, calcium hydroxide or "slaked lime," expanding to several times its original size, with the evolution of much heat. CaO absorbs H_2O from the air, and then CO_2, and gradually crumbles to a coarse powder, becoming "air-slaked lime." *Hydraulic lime* is made from limestone containing more than 10 per cent. of silica, and will harden under water.

Calcium Hydroxide, $Ca(OH)_2$, is slightly soluble in water, and its clear solution is called "lime-water." A film of calcium carbonate will soon form on the surface of lime-water when exposed to the air. Lime-water has an alkaline reaction, *i. e.*, will turn red litmus blue, and acts as a mild alkali.

Uses.—*Whitewash* is a "milk of lime," *i. e.*, lime diffused through water. *Concrete* is a cement of coarse gravel and hydraulic lime. It is of great durability. *Hard finish* is a kind of plaster in which gypsum is used to make the wall smooth and hard. *Calcimine* is a variety of whitewash made of whiting or plaster of Paris. *Mortar* is a mixture of lime and sand wet with H_2O. It hardens by absorbing CO_2 from the air to form a carbonate, and partly, perhaps, by uniting with the SiO_2 of the sand to form a silicate.*

* "If common mortar be protected from the air, it will remain without hardening for many years. It is stated that lime still in the condition of a hydrate has been found in the Pyramids of Egypt. When the ruins of the old castle of Landsberg were removed, a lime-pit, that must have been in existence three hundred years, was found in one of the vaults. The sur-

Lime is valuable as a fertilizer. It acts by rapidly decomposing all vegetable matter, and thus forming NH_3 for the use of plants.* It also sets free the alkalies that are combined in the soil, and furnishes them to the plants, becoming itself a carbonate. Lime is also used extensively in the preparation of bleaching powder, in refining sugar, in making candles, in tanning, and in the manufacture of coal-gas.

Calcium Carbonate, $CaCO_3$, includes limestone, chalk, marble, and marl, and forms the principal part of corals, shells, etc. H_2O charged with CO_2 dissolves $CaCO_3$, which, when the gas escapes on exposure to the air, is deposited. In limestone regions, the water trickling down into caverns has formed "stalactites," which depend from the ceiling, and "stalagmites," that rise from the floor. These frequently assume curious and grotesque forms, as in many limestone caves. Around many springs, the water, charged with $CaCO_3$ in solution, flows over moss or some vegetable substance, upon which the stone is deposited. The spongy rock thus formed is called calcareous tufa, or "petrified moss." (See "Geology," page 49.) *Marble* is crystalline limestone. *Chalk* or *marl* is a porous kind of limestone, formed

face was carbonated to the depth of a few inches, but the lime below this was fresh as if just slaked, and was used in laying the foundations of the new building."—*American Cyclopedia.*

* If applied to a compost heap, it will set free NH_3, thus robbing it of its most valuable constituent. This can be saved by sprinkling the pile with dilute H_2SO_4, or plaster, or by mixing it with dry muck, which will absorb the gas. If there is any copperas (produced by the oxidation of iron pyrites) in the soil, the lime will decompose it, forming gypsum and iron-rust, thus changing a noxious ingredient into an element of fertility.

FIG. 59.

A cave with stalactites and stalagmites.

from beds of shells, but not compressed as in common limestone. *Whiting* is ground chalk.

Calcium Sulphate, ($CaSO_4,2H_2O$), *Gypsum, Plaster, etc.**—This occurs as beautiful fibrous crystals in satin spar, as transparant-plates in selenite, and as a snowy-white solid in alabaster. It is soft, and can be cut into rings, vases, etc. When heated it loses its water of crystallization, and is ground into powder, called "Plaster of Paris," from its abundance near that city. Made into a paste with H_2O, it first swells

* Comparing the formula H_2SO_4 and $CaSO_4$, we see that one atom of Ca can replace two atoms of H; it is therefore one of the class of atoms called **bivalent.**

up, and then immediately hardens into a solid mass. This property fits it for use in copying medals and statues, forming molds, fastening metal tops on glass lamps, etc. *Plaster* (unburned or burned gypsum) is used as a fertilizer.* Its action is probably somewhat like that of lime, and in addition it gathers up ammonia and holds it for the plant.

Calcium Sulphite, $CaSO_3$, should be distinguished from the sulphate. It is much used in preserving cider, being sold as "sulphite of lime."

Calcium Phosphate, "*Phosphate of Lime,*" is frequently termed *bone phosphate*, as it is a constituent of bones. (See p. 113.) It is found in New Jersey, South Carolina,† and Canada. It is the valuable part of certain guanos. Fertilizers are prepared by treating ground bones with H_2SO_4, forming the so-called superphosphate of lime.‡ This is a mixture of gypsum and hydrogen calcium phosphate. The latter

---

* It is said that Franklin brought $CaSO_4$ into use by sowing it over a field of grain on the hill-side, so as to form, in gigantic letters, the sentence, "Effects of gypsum." The rapid growth produced soon brought out the words in bold relief, and decided the destiny of gypsum among farmers.

† "Along the coast of South Carolina are millions of tons of rocks holding this important element of plant-food. The phosphatic beds extend over an area of several hundred square miles, and in some cases they are twelve feet thick. It is estimated that from 500 to 1000 tons underlie each acre." —*Fireside Science.*

‡ Ca_32PO_4 (tricalcium phosphate) $+2H_2SO_4 = H_4Ca2PO_4$ (acid phosphate or superphosphate) $+2CaSO_4$ (calcium sulphate). As the gypsum is only slightly soluble in water, the superphosphate may be removed from the mass by filtering, and used as a fertilizer, or to form phosphorus. In that case it is converted into calcium metaphosphate, $Ca2PO_3$, by evaporating and heating the residue to redness, and then mixed with C and heated again in earthenware retorts. The following reaction takes place:

$$3Ca2PO_3 + 10C = 4P + Ca_32PO_4 + 10CO,$$

being partly changed back into the original form, Ca_32PO_4.

furnishes phosphorus to the growing plant to store in its seeds.—*Example :* corn, wheat.

Calcium Hypochlorite ($CaCl_2O_2$) is an ingredient of chloride of lime or "bleaching powder." This is prepared by passing a current of Cl over pans of freshly slaked lime. It is much used in bleaching and as a disinfectant.

Calcium Chloride, the other compound of bleaching powder, was made in preparing CO_2 (see p. 63). It is used by chemists for drying gases. It absorbs H_2O so greedily that in the open air it will soon dissolve.

STRONTIUM AND BARIUM.

THESE metals are very like Ca. The salts of Ba give a green tint to a flame and those of Sr a beautiful crimson; and are hence much used in pyrotechny. Barium sulphate, commonly called barytes, is found as a white mineral, noted for its weight, whence it is often termed heavy spar. Indeed, the term barium is derived from a Greek word meaning *heavy.* This mineral is largely used for adulterating white-lead. $BaCl_2$ is a test for H_2SO_4. (See p. 106.)

Strontium occurs as *celestine,* $SrSO_4$, and as *strontianite,* $SrCO_3$. Its most important compounds are: the hydroxide, $Sr(OH)_2$, which is used in sugar-making to extract the sugar from molasses; and the nitrate, $Sr2NO_3$, which is used as a constituent of "red fire." The metals themselves are prepared with difficulty, and are of no practical importance.

MAGNESIUM.*

Symbol, Mg.... Atomic Weight, 24...... Specific Gravity, 1.7.

Source.—Mg is found in augite, hornblende, meer-schaum, soap-stone, talc, serpentine, dolomite, and other rocks. Its salts give the bitter taste to sea-water. When pure, it has a silvery luster and appearance. It is very light and flexible. A thin ribbon of the metal will take fire from an ignited match, and burns with a white cloud of MgO, producing such a brilliant light that an ordinary flame casts dense shadows. This light possesses such actinic or chemical properties, that it is used for taking photographs at night, views of coal mines, interiors of dark churches, etc. It has every ray of the spectrum, and so does not, like gas-light, change some of the colors of an object upon which it falls. Magnesium lanterns are occasionally used for purposes of illumination. By means of clockwork, the metal, in the form of a narrow ribbon, is fed in front of a concave mirror, at the focus of which it burns. Mg is prepared from its chloride, $MgCl_2$, by electrolysis, or by reduction by means of Na. It is hoped that the process of manufacture may be cheapened, so that Mg may be brought within the scope of the arts.

---

* With Mg are classed Zn, Cd, and Be. Mg is treated here for convenience, while Zn is described among the useful metals. Cd is used in making some alloys and its salts are employed to some extent in photography and medicine. Be occurs in beryl and emerald. It is of no practical importance.

ALUMINIUM. 145

Compounds.—"*Magnesia alba,*" the common magnesia of the druggist,* is a basic magnesium carbonate. *Magnesium sulphate* ($MgSO_4, 7H_2O$) is known as Epsom salt, from a celebrated spring in England in which it abounds.

———•••———

A L U M I N I U M.†

Symbol, Al.... Atomic Weight, 27.... Specific Gravity, 2.6.

Source.—Al is named from alum, in which it occurs. It is also called the "clay metal." It is the metallic base of clay, mica, slate, and feldspar rocks. Next to O and Si, it is probably the most abundant element of the earth's crust. It is a bright, silver-white metal; does not oxidize in the air, nor tarnish by H_2S. It gives a clear musical ring; is only one fourth as heavy as Ag; is ductile, malleable, and tenacious. It readily dissolves in HCl, and in solutions of the alkalies, but with difficulty in HNO_3 and H_2SO_4. On account of its abundance (every clay-bank is a mine of it) and useful properties, it must ultimately come into common use in the arts and domestic life.

Compounds.—Aluminium Oxide (Al_2O_3).—*Alumina* crystallized in nature, forms valuable Oriental gems.

---

* The magnesia of commerce is made by mixing hot solutions of magnesium sulphate and sodium carbonate. It contains a varying proportion of magnesium hydrate. Dolomite, a rock composed of magnesium carbonate and calcium carbonate, makes a hydraulic cement that "sets" under water.—("Geology," p. 51.)

† Quite a number of rare metals (Ga, In, Sc, etc.) are classed with Al, but none of them are of sufficient interest to be treated in an elementary work.

They are variously colored by the oxides;—blue, in the sapphire; green, in the emerald; yellow, in the topaz; red, in the ruby. Massive, impure alumina is called *emery*, and used for polishing.

Aluminium Silicate ($Al_2O_3 2SiO_2$), *Silicate of Alumina, Common Clay.*—When the clay rocks decay by the resistless and constant action of the air, rain, and frost, they crumble into soil. This contains clay, silica, and also lime, magnesia, oxide of iron, etc. The clay gives firmness to the soil and retains moisture, but is cold and tardy in producing vegetable growth. When free from Fe, it is used for making tobacco-pipes. When colored by ferric oxide, it is known as ocher, and is employed in painting. Common stone and red earthen-ware are made from coarse varieties of clay; porcelain and china-ware require the purest material. Fire-bricks and crucibles are made from a clay which

FIG. 60.

Baking Porcelain.

contains much SiO_2. Fullers' earth is a very porous kind, and by imbibition absorbs grease and oil from cloth.

Glazing.—When any article of earthen-ware has been molded from clay, it is baked. As the ware is porous, and will not hold H_2O, a mixture of the coarse materials from which glass is made is then spread over the vessel, and heated till it melts and forms a glazing upon the clay. Ordinary stone-ware is glazed by simply throwing damp NaCl into the furnace. This volatilizes, and being decomposed by the hot clay makes a sodium silicate over the surface, while fumes of HCl escape. Pb is sometimes used to give a yellowish glaze, which is very injurious, as it will dissolve in vinegar, and form sugar of lead, a deadly poison. The color of pottery-ware and brick is due to the oxide of iron present in the clay. Some varieties have no iron, and so form white ware and brick.

Alum is made by treating clay with H_2SO_4, forming an aluminium sulphate. On adding potassium sulphate a double salt is produced, which separates in beautiful octahedral crystals $(Al_23SO_4,K_2SO_4+24H_2O)$. Instead of the potassium salt, an ammonium salt* is now generally added, and an ammonium alum made, which takes the place of the former in the market.† Alum is much used in dyeing. It unites with the coloring matter, and binds it to the fibers of the cloth. It is therefore called a mordant (*mordeo*, I bite).

SPECTRUM ANALYSIS.

MANY of the metals named as rare have been recently discovered by what is termed Spectrum Analysis. We have already noticed that various

* Ammonium sulphate, from the ammoniacal liquor of the gas-works. (See page 73.)

† There are a large number of other alums known, in which iron, chromium, and manganese are substituted for the aluminium in common alum; all these alums occur in regular octahedra, and can not be separated by crystallization when present in solution together.

metals impart a peculiar color to flame; thus Na gives a yellow tinge; Cu, a green, etc. If now we look at these colored flames through a prism, we shall find, instead of the "spectrum" we are familiar with, a dark space strangely ornamented with bright-tinted lines. Thus the spectrum of Na has one double, yellow line;* K, a violet and a red line; Cs, two beautiful blue lines. Each metal makes a distinctive spectrum, even when the flame is colored by several substances at once. This method of analysis is so delicate that $\frac{1}{180\,000\,000}$ of a grain of Na, or $\frac{1}{1\,000\,000}$ of Li, can be detected in the flame of an alcohol lamp;† while a substance exposed to the air for a moment even will give the Na lines from the dust it gathers. Li has thus been found to exist in tea, tobacco, milk, and blood, although in such minute quantities as to have eluded detection by former methods of analysis.

PRACTICAL QUESTIONS.

1. In the experiment with Na,SO, on page 134, an accurate thermometer will show that in making the solution, the temperature of the liquid will fall, and in its solidification, will rise. Explain.

2. If, in making the solution of Na₂SO,, we use the salt which has effloresced, and so become anhydrous, the temperature will rise instead of falling as before. Explain.

3. Why is KNO₃ used instead of NaNO₃ for making gunpowder?

4. Why is a potassium salt preferable to a sodium one in glass-making?

* The yellow, sodium line consists of two lines lying so closely together as to seem as one. They correspond to Fraunhofer's lines D, as given in the drawings of Kirchhoff and Bunsen.

† For the more perfect examination of the spectra, a "spectroscope" is used. This consists of a tube with a narrow slit at one end, which lets only a narrow beam of light fall upon the prism within, and at the other a small telescope, through which one can look in upon the prism and examine the spectrum of any flame. (See "Astronomy," p. 285.)

5. What is the glassy slag so plentiful about a furnace?

6. State the formulas of niter, saleratus, carbonate and bicarbonate of soda, plaster, pearlash, saltpeter, plaster of Paris, gypsum, carbonate and bicarbonate of potash, sal-soda, and soda.

7. Explain how ammonium carbonate is formed in the process of making coal-gas.

8. Upon what fact depends the formation of stalactites?

9. Why is HF kept in gutta-percha bottles?

10. Explain the use of borax in washing.

11. How are petrifactions formed?

12. In what part of the body, and in what forms, is phosphorus found?

13 Why are matches poisonous? What is the antidote? (See "Physiology," page 209.)

14. Will the burning phosphorus ignite the wood of the match?

15. What principle is illustrated ‧in the ignition of a match by friction?

16. How much H_2O would be required to dissolve a pound of KNO_3?

17. What causes the bad odor after the discharge of a gun?

18. Write in parallel columns (see *Question* 41, page 85), the properties of common and of red phosphorus.

19. What causes the difference between fine and coarse salt?

20. Why do the figures in a glass paper-weight look larger when seen from the top than from the bottom?

21. What is the difference between water-slaked and air-slaked lime?

22. Why do oyster-shells on the grate of a coal-stove prevent the formation of clinkers?

23. How is lime-water made from oyster-shells?

24. Why do newly-plastered walls remain damp so long?

25. Will lime lose its beneficial effect upon a soil after frequent applications?

26. What causes plaster of Paris to harden again after being moistened?

27. What is the difference between sulphate and sulphite of lime?

28. What two classes of rays are contained in the magnesium light?

29. What rare metals would become useful in the arts, if the process of manufacture were cheapened?

30. Why is lime placed in the bottom of a leach-tub?

31. Is saleratus a salt of K or of Na?

32. Why will Na burst into a blaze when thrown on hot water?

33. Why are certain kinds of brick white?

34. Illustrate the power of chemical affinity.

35. Why does not a candle lowered into a jar of Cl go on burning indefinitely?

THE USEFUL METĄLS.

IRON.

Symbol, Fe.... Ątomic Weight, 56.... Specific Gravity, 7.8.

IRON is the symbol of civilization. Its value in the arts can be measured only by the progress of the present age. In its adaptations and employments, it has kept pace with scientific discoveries and improvements, so that the uses of iron may readily indicate the advancement of a nation. It is worth more to the world than all the other metals combined. We could dispense with gold and silver— they largely minister to luxury and refinement—but iron represents solely the results of honest labor. Its use is universal,* and it is fitted alike for massive iron cables, and for screws so tiny that they can be seen only by the microscope, appearing to the naked eye like grains of black sand.

* "Iron vessels cross the ocean,
  Iron engines give them motion,
  Iron needles northward veering,
  Iron tillers vessels steering,
  Iron pipe our gas delivers,
  Iron bridges span our rivers,
  Iron pens are used for writing,
  Iron ink our thoughts inditing,
  Iron stoves for cooking victuals,
  Iron ovens, pots, and kettles,
  Iron horses draw our loads,
  Iron rails compose our roads,
  Iron anchors hold in sands,
  Iron bolts and rods and bands,
  Iron houses, iron malls,
  Iron cannon, iron balls,
  Iron axes, knives, and chains,
  Iron augers, saws, and planes,
  Iron globules in our blood,
  Iron particles in food,
  Iron lightning-rods on spires,
  Iron telegraphic wires,
  Iron hammers, nails, and screws,
  Iron every thing we use."

There is no "California" of iron. Each nation has its own supply. No other material is so enhanced in value by labor.

1 lb. good iron is worth, say..................... $	.04
1 " bar steel...................................	.17
1 " inch-screws..............................	1.00
1 " steel wire.................................	3 to 7.00
1 " sewing-needles.	14.00
1 " fish-hooks...............................	20 to 50.00
1 " jewel-screws for watches..................	3,500.00
1 " hair-springs for American watches	16,000.00*

Source.—Fe is rarely found native, *i. e.*, in the metallic condition. Meteors, however, containing as high as 93 per cent. of Fe associated with Ni and other metals, have fallen to the earth from space. Fe in combination with various other substances is widely diffused. It is found in the ashes of plants and the blood† of animals. Many minerals contain it in considerable quantities. The ores from which it is extracted are generally oxides or carbonates.

Preparation.—Smelting of Iron Ores.—Fe is locked up with O in an apparently useless stone. C is the key that is ready made and left for our use by the Creator. The process adopted at the mines is very simple. A tall blast-furnace is constructed of stone and lined with fire-brick. At the top is the door,

---

* One pound (Troy) of fine gold is worth in standard coin $248.062. All the above statements are based on careful and actual valuation.

† There are only about 100 grains of Fe in the blood of a full-grown person—about enough to make a ten-penny nail—yet it gives energy and life to the system. The metal is often administered as a tonic in the form of citrate or other salt of iron, and is a valuable medicine.

and at the bottom are pipes for forcing in hot air, sometimes twelve thousand cubic feet per minute, by means of pistons driven by steam-power. The

FIG. 61.

A Blast-Furnace.

furnace, after a fire has been started, is filled with limestone, coal (charcoal or coke), and iron ore, in alternate layers. The C* unites with the O of the ore, and goes off as CO and CO_2. The $CaCO_3$ forms

---

* A little N sometimes unites with some C and K, forming potassium cyanide, or with Ti, if any is present, making beautiful copper-colored crystals of titanium cyano-nitride, which are hard enough to scratch glass.

with the SiO_2 and other impurities a richly-colored glassy slag. The iron, as it is reduced, sinks in the molten state to the lowest part of the furnace (the crucible), covered with a layer of slag. The slag runs out in a constant stream from an opening at the proper height, while the iron is drawn off from time to time and run into channels formed in sand. The large main one is called the *sow;* the smaller lateral ones are termed the *pigs*—hence the name *pig-iron.*

Varieties of Iron. — The usual forms are *cast, wrought,* and *steel,* depending upon the proportion of C which they contain. Steel contains more C than wrought iron, and less than cast. The highest proportion of C in cast iron is about 6 per cent., while wrought iron contains only from 0.1—0.7 per cent.

1. **Cast Iron** is the form which comes from the furnace. It is brittle, can not be welded, and is neither malleable nor ductile. It is well adapted for castings, since at the instant of solidification it expands, so as to copy exactly every line of the mold into which it is poured. The castings may be made so soft as to be easily turned and filed, or so hard, by cooling in iron molds,* that no tool will affect them.

2. **Wrought or Malleable Iron** is made by burning the C from cast iron, in a current of highly-heated air, in what is called a reverberatory furnace. The

* These molds are called "chills," and the iron is termed chilled iron. It is used for burglar-proof safes.

iron is stirred constantly, and exposed to the heated air by means of long "puddling-sticks," as they are

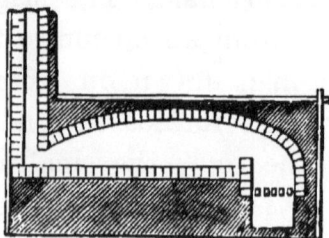

FIG. 62.

A Reverberatory Furnace.

termed. It is taken out while white-hot, and beaten under a trip-hammer to force out the slag; and lastly, pressed between grooved rollers to bring the particles of Fe nearer each other and give it a fibrous structure.* It is now malleable and ductile,† much softer than cast iron, and can be welded.

3. **Steel** contains less C than cast, and more than wrought, iron. It is therefore made from the former by burning out a part of the C, and from the latter by heating in boxes of charcoal, and so adding C.‡ The value of steel depends largely upon its *temper*. This is determined by heating the article, and then allowing it to cool. The higher the temperature the softer the steel. The workman decides this by watching the color of the oxide which forms on the

* This fibrous structure is so noticeable that if a bar of the best Fe be notched with a chisel and then broken by a steady pressure, the fracture will present a stringy appearance, like that of a green stick. By constant jarring, however, Fe tends to take a crystalline structure, becoming rotten and brittle, so that cannon, the axles of cars, etc., are condemned after a certain time, although no flaw may appear.

† It has been beaten into leaves so thin that they have been used for writing-paper—six hundred leaves being only half an inch in thickness—and has been drawn into wire as fine as a hair.

‡ This is termed *case-hardening*. Cheap knives made of soft iron are often covered with a superficial coating of steel in this way. When we use such knives, we soon wear through this crust, and find metal beneath which will take no edge.

surface.* Razors require a straw yellow; table-knives, a purple; springs and swords, a bright blue; and saws, a dark blue tint.†

Bessemer's Process is now extensively used for making steel. Several tons of the best pig-iron are melted, and poured into a large crucible hung on pivots so as to be easily tilted. Hot air driven in from beneath bubbles up through the liquid mass, producing an intense combustion. The roar of the blast, the hot, white flakes of slag ever and anon whirled upward, the long flame streaming out at the top, variegated by tints of different metals, and full of sparks of scintillating iron, all show the play of tremendous chemical forces. The operation lasts about twenty minutes, when the Fe is purified of its C and Si. Enough spiegel-eisen (looking-glass iron), from an ore rich in C and Mn, is added to convert it into steel, when it is poured out and cast into ingots.‡

* The thin pellicles of iron-rust on standing H₂O produce a beautiful iridescent appearance in the same way, the color changing with the thickness of the oxide. Just so a soap-bubble exhibits a play of variegated colors according to the thickness of the film in different parts. (See "Interference of Light," in "Physics.")

† These colors are removed in the subsequent processes of grinding and polishing, but they may be seen in a handful of old watch-springs, to be obtained of any jeweler.

‡ In 1760, there lived at Attercliffe, near Sheffield, a watch-maker named Huntsman. He became dissatisfied with the watch-springs in use. and set himself to the task of making them homogeneous. "If," thought he, "I can melt a piece of steel and cast it into an ingot, its composition should be the same throughout." He succeeded. His steel became famous, and Huntsman's ingots were in universal demand. He did not call them cast steel. That was his secret. The process was wrapped in mystery by every means. The most faithful men were hired. The work was divided, large wages paid, and stringent oaths taken. One midwinter night, as the tall chimneys of the Attercliffe steel-works belched forth their smoke, a belated traveler knocked at the gate. It was bitter cold; the snow fell fast;

Pure Iron.—All varieties of cast iron, wrought iron, and steel, are more or less impure forms of Fe. The pure metal is little known, but can be prepared from its pure salts by reduction or electrolysis.

Compounds.—1. *Black or Magnetic Oxide* (Fe_3O_4) is found in the loadstone, Swedish iron ore, scales which fly off in forging iron, and in mines in various parts of the United States. It is the richest of the ores, and contains as high as 72 per cent. of the metal. 2. *Red Oxide of Iron*, sesquioxide (ferric oxide, Fe_2O_3), is seen in red iron ore, in the beautiful radiated and fibrous specimens of hematite,* specular† iron, red ocher, chalk, bricks and pottery-ware. The sesquioxide, combining with H_2O, forms— 3. *Hydrated Sesquioxide of Iron* (ferric hydroxide, $Fe(OH)_3$). This has a brown or yellow color, which changes to red by heat when the water is expelled, as in the burning of brick, pottery-ware,‡ etc. These

and the wind howled across the moor. The stranger, apparently a common farm-laborer seeking shelter from the storm, awakened no suspicion. The foreman, scanning him closely, at last granted his request, and let him in. Feigning to be worn out with cold and fatigue, the poor fellow sank upon the floor, and was soon seemingly fast asleep. That, however, was far from his intention. Through cautiously opened eyes, he caught glimpses of the mysterious process. He saw workmen cut bars of steel into bits, place them in crucibles, which were then thrust into the furnaces. The fires were urged to their utmost intensity until the steel melted. The workmen, clothed in rags, wet to protect them from the tremendous heat, drew forth the glowing crucibles, and poured their contents into molds. Huntsman's factory had nothing more to disclose. The secret of cast steel was stolen.

* *Hæmatites*, blood-like, from the red color of its powder.
† *Speculum*, a mirror, from the brilliant luster of its steel-gray crystals and mica-like scales in micaceous iron ore.
‡ Clay, containing ferrous oxide (FeO), becomes red by its conversion into ferric oxide.

oxides generally give the brown, yellow, or red tints seen in sand, gravel, etc. The ferric oxide and hydrate are remarkable for the facility with which they absorb O from the air, and impart it to other bodies. This is familiar in the rusting of nails in clapboards, hinges in gate-posts, hooks in ropes, etc., etc.

Iron Carbonate, $FeCO_3$, is found as spathic* and clay iron-stone, and often contains some manganese,† which fits it for the manufacture of certain kinds of steel, whence it is termed steel ore. In chalybeate springs, the free CO_2 in the water holds the $FeCO_3$ in solution. On coming to the air, the CO_2 escapes, and the Fe, absorbing O, is deposited as hydrated ferric oxide, forming the ochry deposit so common around such springs.

Iron Disulphide (FeS_2), *Iron Pyrites, Fool's Gold*— so called, because it is often mistaken by ignorant persons for Au. It occurs in cubical crystals and bright shiny scales. It can be easily tested by roasting on a hot shovel, when we shall catch the well-

---

* *Spath*, spar, as some specimens consist of transparent, shiny crystals, having the same form as calcareous spar (calcium carbonate).

† Manganese is a hard, brittle metal, resembling cast iron in its color and texture. It takes a beautiful polish. Its binoxide, the black oxide of manganese, is used in the manufacture of O, Cl, etc. By fusing MnO_2, $KClO_3$, and KOH. a dark, green mass is obtained called "*chameleon mineral.*" It contains potassium manganate. If a piece of this be placed in H_2O, the solution will undergo a beautiful change from green, through various shades, to purple. This is owing to the gradual formation of permanganic acid. The change may be produced instantaneously by a drop of H_2SO_4. Potassium permanganate is remarkable for the facility with which it parts with its O, and thereby loses its color. It is used extensively as a disinfectant, and as a test of the presence of organic matter. (See page 48.)

known odor of the SO_2. FeS_2 is used as a source of
S, and is roasted to furnish SO_2 in the manufacture
of H_2SO_4.

Ferrous Sulphate ($FeSO_4,7H_2O$), *Green Vitriol,
Copperas*, is made by the action of H_2SO_4 on Fe, and,
at Stafford, Connecticut, and other places, from FeS_2,
by exposure to air and moisture. It is used in dyeing,
making ink, and in photography.

———————•◦•———————

ZINC.

Symbol, Zn.... Atomic Weight, 65.... Specific Gravity, 6.9.
Fusing Point, 811° F. or 433° C.

Source.—Zn is found as ZnO, or red oxide, in New
Jersey, and as ZnS, or zinc blende, in many places.

Fig. 63.

Reduction of Zinc Ore.

Preparation.—ZnO* is smelted
on the same principle as iron ore,
by heating with C. The reaction
is as follows: $ZnO + C = Zn + CO$.

The Zn distils from the crucible
a and is collected in the receiver
d while the CO escapes.

Properties.—Zn is ordinarily
brittle, but when heated to 200°
or 300° F., it becomes malleable,
and can be rolled out into the
sheet Zn in common use. At about
400° F. it is so brittle that it can be powdered in a

* ZnS is roasted to convert it into ZnO.

mortar. It burns in the air with a magnificent bluish light, forming flakes of ZnO, formerly called "Philosopher's Wool." When exposed to the air Zn soon oxidizes, and the thin film of oxide formed over the surface protects it from further change.

Uses.—Its economic uses are familiar. Sheet iron coated with Zn by being dipped in melted Zn forms what is termed *galvanized iron.* Water-pipes lined in this way with Zn are as unsafe as lead (see p. 162) until the Zn is entirely corroded. The oxide and carbonate of zinc are rapidly formed, and these poisonous salts remain in the H_2O. There is the same objection to metallic-lined ice-pitchers. Galvanic action between the metals promotes corrosion. H_2O standing in reservoirs lined with Zn should not be used for drinking purposes. In the case of zinc-covered roofs the rain-water contains zinc.*

Compounds.— **Zinc Oxide,** ZnO, is sold as zinc-white, and is valued as a paint, since it does not blacken by H_2S like white-lead, and is not hurtful to the painter.

Zinc Chloride, $ZnCl_2$, is used as a soldering fluid, which the plumbers prepare by "killing" muriatic acid with Zn. It is also used as a disinfectant and for a number of technical purposes.

Zinc Sulphate, $ZnSO_4,7H_2O$, *White Vitriol*, is used as a "dryer" in oil paints and varnishes. It is decomposed, when strongly heated, into ZnO, SO_2, and O.

---

* When they were first introduced in Boston the washer-women complained that the rain-water was hard, decomposed the soap, and made their hands crack.

T I N.

Symbol, Sn .... Atomic Weight, 118 .... Specific Gravity, 7.3.
Fusing Point, 446° F. or 230° C.

Source.—Sn, though one of the metals longest known to man, is found in but few localities. It is reduced from its dioxide by the action of C.

Properties.—It is soft and not very ductile, but is quite malleable, so that tin-foil is not more than $\frac{1}{1000}$ of an inch in thickness. When quickly bent, a bar of Sn emits a shrill sound, called the "tin cry," caused by the crystals moving upon each other. Sn does not oxidize at ordinary temperatures. Its tendency to crystallize is remarkable.*

Uses.—Common sheet-tin is formed by dipping sheet-iron in melted Sn, which produces an artificial coating of the latter metal. If we leave H_2O in a tin dish, the yellow spots soon betray the presence of Fe. Pins are made of brass wire, upon which a bright white coating of tin is deposited.† Tin is a constituent of a number of important alloys (see p. 177). It forms two classes of salts: the *stannous*, in which it is bivalent, and the *stannic*, in which it is quadrivalent. *E. g.*, $SnCl_2$ and $SnCl_4$.

---

* *Example:* Heat a piece of tin till the coating begins to melt; then cool quickly in H_2O and clean in dilute aqua regia. The surface will be found covered with beautiful crystals of the metal.

† The pins are stuck in papers, as we see them, by machinery which picks them up out of a miscellaneous pile and inserts them in regular rows in the paper, ready for the market. The first part of the process is performed by a sort of coarse comb, which is thrust into the heap, and gathers up a pin in each of the spaces between the teeth.

COPPER.

Symbol, Cu.... Atomic Weight, 63.... Specific Gravity, 8.9.
Fusing Point, about 2012° F. or 1100° C.

Source.—Cu is found native near Lake Superior, frequently in masses of great size. In these mines stone hammers have been discovered, the tools of a people older than the Indians, who probably occupied this continent, and worked the mines. In the western mounds, also, copper instruments are found. The sulphide, *copper pyrites*, is a well-known ore. *Malachite* ($CuCO_3,Cu[OH]_2$), the green carbonate, admits of a high polish, and is made into ornaments of exquisite beauty.

Properties.—Cu is ductile, malleable, and an excellent conductor of heat and electricity.* Its vapor gives a characteristic and beautiful green color to flame. HNO_3 is the solvent of Cu. Its test is NH_4OH, forming in a solution a blue precipitate, which dissolves in an excess of the re-agent with an intense dark blue color.

Compounds.—Copper Acetate, *Verdigris,*† is produced when we soak pickles in brass or copper kettles; the green color which results is caused by this salt—a deadly poison. Preserved fruits, etc., should never stand in such vessels, as the vegetable acids dissolve Cu readily.

---

* Commercial Cu is never quite pure. Its properties are affected very markedly by the pressure of even minute amounts of foreign substances.

† The term verdigris is sometimes incorrectly applied to the green coating of carbonate, which gathers upon brass or copper in a damp atmosphere.

Copper Oxide, CuO, is the black coating which is formed on copper or brass kettles. It dissolves readily in fats and oils. Such utensils should therefore be used only when perfectly bright, and never with fruits, sweetmeats, jellies, pickles, etc.

Copper Sulphate ($CuSO_4,5H_2O$), *Blue Vitriol,* is much used in dyeing, calico printing, and in voltaic batteries.

LEAD.

Symbol, Pb.... Atomic Weight, 207...... Specific Gravity, 11.36.
Fusing Point, 635° F. or 335° C.

Source.—The most common ore of Pb is galena, PbS, which is reduced by processes which differ according to the purity of the ore.

Properties.—Pb is malleable; but contracts as it solidifies, so it can not be used for castings. Lead itself is not poisonous, and " bullets have been swallowed, and then thrown off without any harm except the fright." The soluble salts of Pb are, however, all very poisonous. The effects seem to accumulate in the system, and finally to manifest themselves in disease. Persons who work in lead compounds, as painters, after a time suffer with colics, paralysis, etc., while plumbers, who handle only metallic Pb, do not suffer.

Uses.—Pb is much used for water-pipes, and is the most convenient of any metal for that purpose. Pure H_2O passing through the pipe will not corrode the Pb, but the O of the air it contains forms an

oxide of lead which dissolves in the H_2O, leaving a
fresh surface for oxidation. If there are any sulphates or carbonates in the H_2O, they will form a
coating over the Pb, and protect it from further
corrosion; and as carbonate of lime is common in
hard water, that is generally safe. If, when we
examine a lead pipe that is in constant use, we find
it covered with a white film, it is a good sign; but
if it is bright, there is cause for alarm. Still, however much may be said about the danger, people will
use lead pipes, and the following precautions should
be observed: *Before using the water in the morning,
always let it run long enough to remove all which has
remained in the water-pipes during the night;* and
when the H_2O is let on again after it has been shut
off for awhile, *leave the faucet open until the pipe
is thoroughly washed.*

The Test of Pb is H_2S, forming lead sulphide,
PbS. The following is an interesting illustration:
Thicken a solution of lead acetate with a little gum-
arabic, so as not to flow too readily from the pen,
and then make any sketch which your fancy may
suggest. This, when dry, will be invisible. When
it is to be used, dampen the paper slightly on the
wrong side, and then direct against it a jet of H_2S.
The picture will at once blacken into distinctness.*

Compounds.—Lead Oxide, PbO, the well-known
litharge, is formed by heating Pb in a current of

---

* A delicate test for the presence of lead in water is this: Add a few
drops of acetic acid and then a small pinch of powdered bichromate of
potassium ($K_2Cr_2O_7$). If Pb is present, a yellow turbidity will appear.

air.* *Lead Dioxide*, PbO_2, is formed by oxidizing PbO. A mixture of the two, called *red-lead*, is used for coloring red sealing-wax, and as a paint.

Lead Carbonate ($PbCO_3$), *White-Lead*, consists of basic lead carbonates, and is made as follows:

FIG. 64.

A.—*An earthen pot.*
L.—*A coil of lead.*
V.—*A solution of vinegar.*

Thousands of earthen pots fitted with covers and containing weak vinegar (acetic acid) and a small roll of Pb, are arranged in piles, and then covered with tan-bark. The acetic acid combines with the Pb, but the CO_2 formed by the decomposing tan-bark creeps in under the cover, driving off the acetic acid, and forming lead carbonate. The acetic acid, thus dispossessed, attacks another portion of the Pb, but is robbed again; and so the process goes on, till the Pb is exhausted. White-lead is often adulterated with heavy spar, gypsum, etc.

Lead Acetate, *Sugar of Lead*, has a sweet, pleasant taste, but is a virulent poison. Its antidote is Epsom salt, which forms an insoluble lead sulphate. H_2O dissolves sugar of lead readily. If a piece of Zn, cut in small strips, be suspended in a bottle filled with a solution of lead acetate, the Pb will be deposited upon it by voltaic action in beautiful metallic spangles, forming the "lead-tree."

FIG. 65.

The Lead-tree.

---

* *Example*: Heat a bit of lead upon charcoal in the oxidizing flame of the blow-pipe. A film of the suboxide forms first, then a yellow crust of the oxide.

THE NOBLE METALS.

Au, Ag, Pt, Hg, Pd, Ir, Os, Ru, aŋd Ro.

GOLD.

Symbol, Au.... Atomic Weight, 196.2.... Specific Gravity, 19.34.
Fusing Point, about 2012° F. or 1100° C.

Sources.—Au is sometimes found in masses called nuggets, but generally in scattered grains, or scales. As the rocks in which it occurs disintegrate by the action of the elements and form soil, the Au is gradually washed into the valleys below, and thence into the streams and rivers where, owing to its specific gravity, it settles and collects in the mud and gravel of their beds.*

Preparation.—As the metal is thus found native, the process is purely mechanical, and consists simply in washing out the dirt and gravel in wash-pans, rockers, sluices,† etc., at the bottom of which the Au accumulates. In the quartz-mills, the rock is thrown into troughs of water where by heavy stamps the

* In California, Au is found in the detritus (small particles of rock worn off by attrition) of granite and quartz. It occurs in the gravel of hills from the surface to the "bed-rock," sometimes a depth of 300 to 500 feet; in the alluvial soil of the plains, and even in vegetable loam among the roots of grass.

† Sluices are generally used in California. These are gently inclined troughs, sometimes extending for miles. Across the bottom are fastened low wooden bars, called *riffles*, above which quicksilver is placed. The dirt is shoveled into these sluices, or the auriferous hills are cut down, dissolved, and washed through them by powerful streams of water, which are constantly running. The H_2O floats off the debris, while the Hg catches the gold.

ore is crushed to powder. As the thin liquid mud thus formed splashes up on either side, it runs over broad, metallic tables covered with Hg; or is washed through a fine wire-screen, and carried to the "amalgamating-pans" by a little stream of water. The Hg unites with the particles of Au and forms with them an *amalgam* (a compound of mercury and a metal). Hg is easily separated from Au by distillation,* and collected to be used again.

Quartation.—Au is commonly found alloyed with Ag. The Ag is then dissolved out by HNO_3. There must be at least three parts of Ag to one of Au, else the gold will protect the silver from the action of the acid. If there is not so much, some is fused with the alloy.†

Properties.—Pure Au is nearly as soft as Pb. It is extremely malleable ‡ and ductile. Its solvent is

---

* The larger part of the Hg is separated from the amalgam by pressure in canvas or buckskin bags, the liquid Hg escaping through the pores, while the amalgam is left quite dry. The latter is then "retorted" for distillation.

† "In works for the refining of gold and silver, the processes can be conducted economically only when great care is taken to avoid the loss of any particles of the precious metals. Thus all the old crucibles are ground and treated with mercury, and after as much gold and silver as possible have been extracted, the residues are sold to the *sweep-washers*, who extract a little more by melting with lead. The very dust off the floors is collected and treated in a similar way."—BLOXAM.

‡ For a description of the process of making gold-leaf, see "Physics," page 20. "When one of these leaves is held up to the light, it exhibits a beautiful green color, and if it be rendered still thinner, either by beating, or by floating it upon a very weak solution of potassium cyanide, which slowly dissolves it, it transmits, when taken upon a glass plate and held up to the light, a blue, violet, or red light, in proportion as its thickness diminishes. Even when it is so transparent that one may read through it, the yellow color and luster of the gold are still visible by reflected light. These varying colors of finely-divided gold are turned to account in the coloring of glass and in painting on porcelain."

aqua regia. It does not oxidize at any temperature and, on account of the resistance it offers to corrosion, it was anciently called the king of the metals.

S I L V E R .

Symbol, Ag... Atomic Weight, 107.7... Specific Gravity, 10.57.
Fusing Point, 1904° F. or 1040° C.

Sources.—Silver is found throughout the West in a great variety of forms—most commonly, however,

Fᴵɢ. 66.

Separation of Pb *from* Ag. (*See Bloxam's Metals.*)

combined with S, as *black sulphide*, Ag_2S; with Cl, forming *horn-silver*, AgCl; with S and As or Sb, making *ruby-silver*, and also associated with P̂b in ordinary galena.

Preparation.—1st. The *sulphide* is treated as follows: The ore is crushed into fine powder and then roasted with common salt. The Cl of the salt unites with the Ag, forming silver chloride. This is next put into a revolving cylinder with H_2O, Hg, and iron scraps. The Fe removes the Cl from the silver, when the Hg takes it up, thus forming an amalgam of Hg and Ag. From this the Ag is easily obtained by distilling off the Hg, as in the extraction of gold.* 2d. From *horn-silver*, AgCl, the process is like the latter part of that just described. 3d. From *lead* the Ag can be profitably obtained when there are only two or three ounces in a ton. The alloy of the two metals is melted, and then slowly cooled. Crystals of almost pure Pb appear, and are skimmed out as fast

FIG. 67.

A Cupel.

as formed, thus leaving an alloy much richer in silver. (See Fig. 66.) The last portions of Pb are removed from this alloy by " cupellation."

Cupellation.—A cupel (*cupella*, a small cup) is a shallow vessel, made of bone ashes. In this the Ag, debased with Pb and other impurities, is exposed to a red heat, so as to melt the metals, while a current of hot air plays

* " The process of reducing silver ores at the West is unlike the German method given above, and varies in different localities. One plan is as follows: The powdered and roasted Ag_2S is placed with Hg in iron pans, five feet in diameter and two feet deep. Here it is kept heated by steam to 180°, and agitated by revolving stirrers. The chloride is not roasted, but is simply powdered, and then worked in the pans for an hour with NaCl before adding the Hg."—STEVENSON.

upon the surface. The Pb oxidizes to PbO, and is absorbed by the porous cupel. The mass appears soiled and tarnished, but the refiner keeps his eye upon it as the process continues, watching eagerly, until at last there is a brilliant play of colors—in a

Fig. 68.

Cupels in the Furnace.

moment more the last film of oxide disappears, and the brilliant surface of the pure silver lies gleaming at the bottom.*

* See Malachi iii. 3. During the cooling of the cake of Ag, some very remarkable phenomona are observed. When a thin crust of metal has formed upon the surface, the Ag beneath it assumes the appearance of boiling, and the crust is forced up into hollow cones about an inch high, through which the melted Ag is thrown out with explosive violence, some of it being splashed against the arch of the furnace, and some solidifying into most fantastic tree-like forms several inches in height. This behavior of Ag has been shown to be due to its property of mechanically absorbing O, at a temperature above its melting-point, which it gives off as it approaches the point of solidification, the escaping gas forcing up the crust of solid Ag formed upon the surface.

Properties.—Ag is the whitest of the metals. It is malleable and ductile, and is of all the metals the best conductor of electricity. It expands at the moment of solidification, and therefore can be cast. It has a powerful attraction for S, forming silver sulphide. Silver spoons and door-knobs are tarnished by the minute quantities of H_2S present in the air.* The best solvent of Ag is HNO_3. The test of Ag in solution is HCl, which forms a cloudy precipitate of silver chloride. A solution of silver coin is blue, from the Cu it contains. Standard silver is whitened by being heated until the O of the air has converted a little of the Cu on the outside into CuO, which is dissolved by immersing in dilute H_2SO_4 or NH_4OH. The film of nearly pure Ag which then remains at the surface exhibits a want of luster and is called *dead* or *frosted silver.* It is brightened by burnishing.

Compounds.—**Silver Nitrate,** $AgNO_3$, is sold in small, round sticks as *lunar caustic,* used as a cautery. It stains the skin and all organic matter black, especially when exposed to the light, owing to the formation of metallic silver.† Many hair-dyes and indelible inks contain $AgNO_3$. It is also the basis

---

* "Those who have visited sulphur springs know the propriety of carefully protecting their watches, and of never wearing silver ornaments to the hot baths. Ag_2S is very easily dissolved by a little *dilute* ammonia (1 part of NH_4OH to 20 of H_2O), which is therefore used for cleaning silver door-knobs.—*Oxidized silver*, as it is erroneously called, is made by immersing articles of silver in a solution obtained by boiling sulphur with potash, when the metal becomes coated with a thin film of sulphuret of silver."— BLOXAM.

† The stain of silver nitrate may be removed by a strong solution of potassium iodide or the poisonous potassium cyanide. (See caution on page 290.)

of photography (light-drawing) and daguerreotyping,* which are both founded upon essentially the same principles. The general outlines of the photographic process are as follows: 1. Iodized collodion† is poured upon a clean glass plate, which, on evaporation, it covers with a transparent film. 2. The plate is put in the "nitrate of silver bath,"‡ where the salt of silver is absorbed by the collodion film and changed to brom-iodide of silver. The plate is now ready for the picture. After the sitting, the plate is taken, carefully protected from the light, to the operator's

* The daguerreotype is named from M. Daguerre, the discoverer, who received a pension of 6,000 francs per year from the French government. A plate of Cu, plated on one side with Ag, is exposed to the vapor of I and Br until a compound of brom-iodide of silver is formed upon the surface. This is extremely sensitive to the light, hence the process is always conducted in a dark closet. The plate is then carried, carefully covered, to the camera, and placed in the focus, where the rays of light from the person whose "picture is being taken" fall directly upon it. These rays decompose the brom-iodide of silver. The amount of this change is directly proportional to the intensity of the light that is reflected from different parts of the person to form the image in the camera. A white garment reflects much of the light that falls upon it, so the corresponding part of the plate will be very much changed. A black garment reflects but little light, so that part will not be changed at all. The different colors and shades reflect varying proportions of light, and so influence the plate correspondingly. When the plate is taken out of the camera, it is carefully covered again and carried into the dark closet. No change can be detected by the eye; but on exposure to the vapor of Hg, wherever the Ag has been freed, the Hg will combine with it, forming a whitish amalgam, but it has no effect on the rest of the plate. The picture thus treated comes forth nearly perfect in its lights and shades. The undecomposed brom-iodide of silver is removed by a solution of sodium hyposulphite. A solution of gold chloride and sodium hyposulphite is then poured upon the plate and warmed. This golden varnish finishes the picture.

† Iodized collodion is composed of gun-cotton dissolved in alcohol and ether, to which are added ammonium iodide and cadmium bromide, or similar salts.

‡ The nitrate of silver bath contains nitrate of silver and iodide of silver in solution, and is acidulated with nitric acid.

room. Here the picture is, 3, *developed* by a solution
of ferrous sulphate (green vitriol, see p. 158) or pyrogallic acid (see p. 238) ; at the right stage the liquid
is washed off, and the operation checked. 4. It is
fixed with a solution of sodium hyposulphite, which
dissolves the unaltered brom-iodide of silver. 5. It is
washed, dried, and coated with amber varnish to
preserve the film from accidental injury. The
"*negative*" is now completed, and is a correct likeness, only the lights and shades are reversed. From
this the pictures are, 1, "printed" by placing the
negative upon a sheet of prepared paper,* and exposing it to the sun's rays. When the colors are sufficiently deepened, the picture is, 2, *toned* in the
"toning-bath," which contains a little "bicarbonate
of soda" and a minute quantity of gold chloride ;
3, *fixed*, by sodium hyposulphite which dissolves the
unaltered AgCl ; 4, thoroughly washed in water frequently renewed ; and, lastly, dried and mounted on
card-board. The thoroughness of the third and fourth
processes has much to do with the permanence of
the picture. If any of the chloride or the compound
formed by the hyposulphite be left, it will cause
fading or discoloration.

Silver Chloride, AgCl, occurs naturally as horn-
silver, and falls as a white curdy precipitate when
HCl or a soluble chloride (*e. g.*, NaCl) is mixed with a
silver solution.

---

* This paper is "sensitized" by floating it on a solution of sodium
chloride, and then on one of silver nitrate, thus filling the pores of the
paper with the silver chloride, which is extremely sensitive to light.

PLATINUM.

Symbol, Pt.... Atomic Weight, 194.2.... Specific Gravity, 21.53. Fusing Point, about 3632° F. or 2000° C.

Source.—Pt* is chiefly found in the Ural Mountains, where it occurs in alluvial deposits, usually in small, flattened grains,† which contain Au, Pd, Rh, Ru, Ir, and Os, as well as Fe, Cu, etc., besides the Pt.

Preparation.—The "ore," as it is called, is separated from the earthy particles by washing, and the Pt extracted by a rather complicated process.

Properties.—Pt resembles Ag in its appearance. It is one of the most ductile metals, wire having been made from it so fine as to be invisible to the naked eye.‡ It is soluble in aqua regia, but not in the simple acids. It does not oxidize in the air, is one of the most infusible of metals, and can be melted only by the heat of the compound blow-pipe or voltaic battery. In the arts it is fused in the former manner. These properties fit it for making crucibles that are invaluable to the chemist.

Platinum Sponge (see page 42) is made by heating the double chloride of Pt and NH₄.

Platinum Black is obtained by the action of reducing agents upon Pt solutions.

---

* The word *platinum* signifies "little silver."

† The largest nugget ever found weighed about 18 lbs.

‡ Wollaston's Method, as it is called, consists in covering fine platinum wire with several times its weight of Ag, and then drawing this through the plates used for drawing wire until the finest hole is reached, when the wire is placed in HNO₃, which dissolves the Ag and leaves the Pt intact. This, in the form of the finest wire known, may be found in the solution by means of a microscope. (See "Physics," p. 19.)

MERCURY.

Symbol, Hg.... Atomic Weight, 200.... Specific Gravity, 13.6.
Melting (Freezing) Point, —39° F.... Boiling Point, 680° F. or 360° C.

MERCURY is also called quicksilver, because it rolls about as if it were alive, and was supposed by the alchemists to contain silver. It was known very anciently, and the mines of Spain were worked by the Romans.

Source.—Cinnabar, HgS, a brilliant red ore, is the principal source of this metal.*

Preparation.—Hg is readily prepared by heating HgS in a current of air. The S passes off as SO_2, while the Hg volatilizes and is condensed in earthen pipes.

Properties.—Hg emits a vapor at all temperatures, and this vapor is poisonous. The solvent of Hg is HNO_3. It forms an amalgam † with gold or silver. This is its most singular property. A gold leaf

---

* Hg is found native in Mexico in very small quantities, where the mines are said to have been discovered by a slave, who, in climbing a mountain, came to a very steep ascent. To aid him in surmounting this, he tried to draw himself up by a bush which grew in a crevice above. The shrub, however, giving way, was torn up by the roots, and a tiny stream, of what seemed liquid silver, trickled down upon him.

† "Several years ago, while lecturing upon chemistry before a class of ladies, we had occasion to purify some quicksilver by forcing it through chamois skin. The scrap of leather remained upon the table after the lecture, and an old lady, thinking it would be very nice to wrap her gold spectacles in, accordingly appropriated it to this purpose. The next morning she came to us in great alarm, stating that the gold had mysteriously disappeared, and nothing was left in the parcel but the glasses. Sure enough, the metal remaining in the pores of the leather had amalgamated with the gold, and, entering, destroyed the spectacles. It was a mystery, however, which we could never explain to her satisfaction."—J. R. NICHOLS in Fireside Science,

dropped upon mercury disappears like a snow-flake in water. Particles of Ag or Au, too fine to be seen by the eye, will be found by Hg and gathered from a mass of ore.

Uses.—Hg is extensively employed in the manufacture of thermometers and barometers; for silvering mirrors;* and for extracting the precious metals from their ores.

The action of Hg on the human system is too well known to need description. "In its metallic state, Hg has been taken with impunity in quantities of a pound weight" ("American Cyclopedia"), but when finely divided, as in vapor, mercurial ointment,† or "blue-pill," its effects are marked. It renders the patient extremely susceptible to colds; acts, as is generally thought, upon the liver, increasing the secretion of bile, and repeated doses produce "salivation."

Compounds.—**Mercuric Oxide,** HgO, "red precipi-

---

* Mirrors were anciently made of steel or silver, highly polished. They were very liable to rust and tarnish, and so a piece of sponge, sprinkled with pumice-stone, was suspended from the handle for rubbing the mirror before use. Seneca, in lamenting over the extravagance of his time among the old Romans, says: "Every young woman nowadays must have a silver mirror." The process of "silvering" ordinary mirrors is briefly as follows: Tin-foil is first spread evenly upon a marble table, and then the Hg is carefully poured over it. The two metals combine, forming a bright amalgam. A clean, dry plate of glass is then carefully pushed forward over the table so as to carry the superfluous Hg before it, and also prevent the air from getting between the glass and the amalgam. Weights are afterward added to cause the film to cling more closely. In twenty-four hours the plate is removed, and in three or four weeks is dry enough to be framed. When we look in a mirror we rarely realize what it has cost others to thus minister to our comfort. The workmen are short-lived. A paralysis sometimes attacks them within a few weeks after they enter the manufactory, and it is thought remarkable if a man escapes for a year or two.

† This is vulgarly called "anguintum," which may be a corruption of the Latin term unguentum (unguent). It is used in cutaneous diseases.

tate," is interesting, as the substance from which Priestley discovered O gas.

Mercurous Chloride, HgCl, *Calomel*, is a white powder used in medicine. It can be easily distinguished from corrosive sublimate, since it is insoluble in H₂O, and hence tasteless.

Mercuric Chloride, HgCl₂, *Corrosive Sublimate*, is a heavy, white solid, soluble in H₂O, and has a burning metallic taste. It has powerful antiseptic properties, and is used to preserve specimens in natural history. It is a deadly poison. Its antidote is white of eggs, milk, etc.

Mercuric Sulphide, HgS, "*Vermilion*," is made by subliming a previously fused mixture of Hg and S. It is used as a pigment.

THE ALLOYS.

THESE are very numerous, and many of them possess properties so different from their elements that they seem like new metals. The color and hardness are changed, and sometimes the melting point is below that of any one of the constituents. The proportions of the metals used vary. The following is a fair average:

Type-metal* contains 50 per cent. of Pb, equal parts of Sn and antimony,† and a little Cu.

---

* The composition of type-metal varies considerably. It is sometimes made of Pb and antimony alone; and the proportions of these two metals are different for large and small type—the small type containing a larger amount of antimony to make them harder.

† Antimony was discovered by Basil Valentine, a monk of Germany, in

Pewter contains 9 parts of Sn and 1 of Pb.

Britannia consists of 9 parts of Sn, 1 of Sb, and usually about 3 per cent. Zn, and 1 per cent. Cu.

Brass is about 2 parts of Cu and 1 of Zn.

German Silver contains 50 parts of Cu, 25 of Zn, and 25 of Ni* (brass whitened by nickel).

Soft Solder, used by tinsmiths, is made by melting Pb and Sn together, the usual proportion being half-and-half.

Hard Solder is composed of Cu and Zn.

Wood's Fusible Metal melts at 158° F.; and spoons made of it will fuse in hot tea. It can be melted in a paper crucible over a candle. It consists of Bi,† Pb, Sn, and Cd. Yet the first metal melts at 518° F., the second at 635°, the third at 446°, and the fourth at 599°.

Bronze is 3 parts of Cu and 1 part of Sn.

Gold is soldered with an alloy of itself and Ag; **Silver**, with itself and Cu; **Copper**, with itself and

the fifteenth century. It is a brittle, bluish-white metal, with a beautiful laminated, star-like, crystalline structure. Its chief use is as an alloy for type-metal, Britannia-ware, etc., one of its most important properties being that it expands in cooling from fusion, and thus makes a very sharp casting. Its test is H_2S, which throws down a brilliant orange-colored precipitate. Melt a small fragment of Sb before the blow-pipe, and throw the melted globule upon an inclined plane. It will instantly dart off in minute spheres, each followed by a long trail of smoke.

* Ni, like Co, is a constituent of meteorites. It is mined in Pennsylvania for the United States Government to make into coins. Formerly, its principal use was in German silver, but of late it has been employed extensively in the manufacture of the best plated-ware. (See "Physics," page 238.) Its silvery whiteness, when pure, its high polish, which often lasts for years, and its hardness, almost equal to that of steel, eminently fit it for the plating of mathematical and other delicate instruments. The salts of Ni have a handsome green tint.

† Bismuth closely resembles Sb. Its chief use is in making fusible alloys. Its salts are employed in making cosmetics and as medicines.

Zn: the principle being that the metal of lower fusing point causes the other to melt more easily.

Coin.—The precious metals, when pure, are too soft for common use. They are therefore hardened by other metals. The gold coin of the United States consists of 9 parts of gold and 1 of alloy. The alloy is chiefly Cu; but gold coin always contains a little Ag, which was not separated from the Au in the quartation (p. 166). Silver coin is 9 parts of Ag and 1 of Cu. The nickel coin is 75 parts of Cu and 25 of Ni. Cu being cheaper than Ni, it is used to make the coin larger. The term *carat*, applied to the precious metals, means $\frac{1}{24}$ part. Therefore, gold 18 carats fine contains $\frac{18}{24}$ of gold and $\frac{6}{24}$ of alloy.

Shot is an alloy of something less than 1 part of As to 100 of Pb. The manufacture is carried on in what are called "shot-towers," some of which are two hundred and fifty feet high. The alloy is melted at the top of the building, and poured through colanders. The metal, in falling, breaks up into drops, which take the spherical form (see "Physics," pages 44 and 192), harden, and are caught at the bottom in a well of water, which cools the shot and also prevents their being bruised in striking. The shot are dipped out, dried, and then assorted, by sifting in a revolving cylinder, which is set slightly inclined and peforated with holes, increasing in size from the top to the bottom. The shot being poured in at the top, the small ones drop through first, next the larger, and so on, till the largest reach the bottom. Each size is received in its own box. Shot are polished by being agitated for several hours with black-lead, in a rapidly-revolving wheel. They are finally tested by rolling them down a series of inclined planes placed at a little distance from each other. The spherical shot will jump from one plane to the next, while the imperfect ones will fall short, and drop below; or sometimes, by rolling down a single inclined plane, the spherical ones will go to the bottom, while the imperfect ones roll off at the sides.

Or-Molu is a beautiful alloy of Cu and Zn resembling red gold, but it soon tarnishes by exposure to the air.

Aluminium Bronze, or gold, is an alloy of Al and Cu. It is elastic, malleable, and very light. It strikingly resembles gold, and is sometimes used instead of that costly metal.

REVIEW OF THE PROPERTIES OF THE METALS.

Oxidation.—K and Na have an intense attraction for O and other elements, and are never found except in combination, while Au, Pt, etc., have little affinity for other substances, and are therefore found native.

Density.—Li is lighter than any known liquid. K, Na, and Li float upon H_2O, while Pt is over twenty-one times and Os over twenty-two times as heavy as H_2O.

Melting Point.—Hg is liquid at all ordinary temperatures. K and Na melt beneath the boiling point of H_2O; Zn below a red heat, and Cu above; Co, Ni, and wrought iron require the greatest heat of the forge (4000° F.), while Pt and Os melt only in the flame of the oxy-hydrogen blow-pipe. Sn melts at the lowest temperature (446°) of any of the ordinary metals.

Color.—The most common color is white, of varying shades. It is nearly pure in Ag, Pt, Cd, and Mg; yellowish in Sn; bluish in Zn and Pb; gray in Fe, and reddish in Bi. Cu is a full red, and Au a bright yellow.

Malleability.—Au, Ag, Al, and Cu are the most malleable of the metals; Au, Ag, and Pt are the most ductile.

Brittleness.—Sb and Bi may be easily powdered; Zn may be broken with more difficulty, while the fibrous metals are exceedingly tough.

Tenacity.—Steel is the most, and lead the least, tenacious of the metals; the proportion being as 1 to 42.

Special Properties.—Certain of the metals are valuable because of their peculiar properties. Thus, Hg, because it will form an amalgam, and is a liquid at all ordinary temperatures; Sb, because it hardens Pb and Sn; Bi and Cd, because they render Pb and Sn more fusible; Ni, because it whitens Cu; Mg, for its brilliant light; Au, for its rarity and luster; Fe, for the diverse properties it can assume in wrought and cast iron and in steel, and because it is the only metal which can be used for the magnetic needle and electro-magnet; Cu, for its ductility and its conductivity of electricity; and Pt, for its infusibility.

PRACTICAL QUESTIONS.

1. Pb is softer than Fe; why is it not more malleable?

2. What is the cause of the changing color often seen in the scum on standing water?

3. How can the spectra of the metals be obtained?

4. Ought cannon, car-axles, etc., to be used until they break or wear out?

5. Why is "chilled iron" used for safes?

6. Does a blacksmith plunge his work into wa er merely to cool it?

7. What causes the white coating made when we spill water on zinc?

8. Is it well to scald pickles, make sweetmeats, or fry cakes in a brass kettle?

9. What danger is there in the use of lead pipes? Is a lining of Zn or Sn a protection?

10. Is water which has stood in a metal-lined ice-pitcher healthful?

11. If you ask for "cobalt" at a drug-store, what will you get? If for "arsenic"?

12. What two elements are fluid at ordinary temperatures?

13. Should we touch a gold ring to mercury?*

14. Why does silver blacken if exposed to the air?

15. Why does silver tarnish more rapidly where coal is used for fires?

16. Why is a solution of a silver coin blue?

17. Why will a solution of silver nitrate curdle brine?

18. Why does writing with indelible ink turn black when exposed to the sun, or to a hot iron?

19. What alloys resemble gold?

20. Why does a fish-hook "rust out" the line to which it is fastened?

21. Why do the nails in clapboards loosen?

22. Show that the earth's crust is mainly composed of burnt metals.

23. What kind of iron is used for an electro-magnet? For a magnetic needle?

24. Why does a "tin" pail so quickly rust out when once the tin is worn through?

25. Why is the zinc oxide found in New Jersey red, when zinc rust is white?

26. Should we filter a solution of permanganate of potash through paper?

27. Why are wood, cordage, etc., sometimes soaked in a solution of corrosive sublimate?

28. Why does the white paint around a sink sometimes turn black? What danger does this indicate?

* If the surface is only whitened, the Hg may be removed with dilute HNO_3, and the ring be polished to look as before. The Hg will soon penetrate the gold, and render it brittle.

29. Why is aluminium, rather than platinum, sometimes used for making the smallest weights?

30. How would you detect the presence of iron particles in black sand?

31. Which metals can be welded?

32. When the glassy slag from a blast-furnace has a dark color, what does it show?

33. In welding iron, the surfaces to be joined are sometimes sprinkled with sand. Explain.

34. What is the difference between an alloy and an amalgam?

35. Steel articles are blued to protect from rusting, by heating in a sand-bath. Explain.

36. Give the formulas for copperas and white lead.

37. Why is Hg used for filling thermometers?

38. What oxides are formed by the combustion of Na, K, Zn, S, Fe, Pb, Cu, P, etc? Which are bases? Anhydrides? Give the common name of each.

39. Is charcoal lighter than H_2O?

40. Name the "vitriols."

41. Is Mg univalent or bivalent? Zn?

42. Name some bibasic acid.

43. Name a neutral salt. An acid salt.

44. Calculate the percentage of water contained in crystallized copper sulphate; in sodium sulphate; in calcium sulphate; in alum.

45. What is the test for Ag? Cu?

46. What weight of crystallized "tin salts" ($SnCl_2,2H_2O$) can be prepared from one ton of metallic tin?

47. 100 parts by weight of silver yield 132.87 parts of silver chloride. Given the atomic weight of chlorine (35.4), required that of silver.

48. What is the composition of slaked lime?

49. How is ferrous sulphate obtained? How many tons of crystals can be obtained by the slow oxidation of 230 tons of iron pyrites containing 37.5 per cent. of sulphur?

50. Required 500 tons of soda crystals; what will be the weight of salt and pure sulphuric acid needed?

51. Describe the uses of lime in agriculture.

52. How many tons of oil of vitriol, containing 70 per cent. of pure acid (H_2SO_4), can be prepared from 250 tons of iron pyrites, containing 42 per cent. of sulphur?

III.

ORGANIC CHEMISTRY.

―――――•♦•―――――

" Thus the Seer,
  With vision clear,
  Sees forms appear and disappear,
  In the perpetual round of strange,
  Mysterious change
  From birth to death, from death to birth,
  From earth to heaven, from heaven to earth
  Till glimpses more sublime
  Of things, unseen before,
  Unto his wondering eyes reveal
  The Universe as an immeasurable wheel
  Turning forevermore
  In the rapid and rushing river of Time."

LONGFELLOW.

ORGANIC CHEMISTRY.

INTRODUCTION.

Organic Chemistry is the chemistry of the compounds of carbon. It originally meant the chemistry of organized bodies, animal and vegetable, and their products. As was stated in the introduction (p. 7), it was formerly believed that the so-called organic substances formed a group entirely distinct from the inorganic, because they could be produced only by the agency of life. But in 1828, Wöhler, a German chemist, discovered that an organic compound, urea, could be artificially prepared. Since then a very large number of compounds have been made, which had before been obtained only from animal and vegetable substances; among others such well-known ones as alcohol, tartaric acid, glycerin, indigo, alizarin (the coloring matter of madder), and gallic acid.*

There is, however, one class of organic substances, the *organized* bodies, such as muscular tissue, nerve structure, ligneous fiber, no one of which has yet been made in the chemist's laboratory; nor is there any promise in the past development of organic chemistry, wonderful as that development has been,

---

* "There is but little doubt, as new methods are discovered, and our knowledge of the carbon compounds increase, that we may eventually be able to produce synthetically even the most complex."—MILLER.

that these substances can be artificially made. These organized bodies are formed from inanimate matter, by the action of life, and are constantly undergoing rapid change.

While other substances are formed and remain fixed in one state under the influence of chemical affinity, the organized bodies grow rapidly, change constantly, and when life ceases, as rapidly decay. Owing to their complex structure, and the presence in many of them of the negative N, they form most unstable compounds. In this we find the cause of their quick decay. The vital principle alone holds them together, frequently in opposition to the laws of chemical affinity; and the instant that is removed, the tendency is to seek new affinities and form new compounds.

Composition of Organic Substances.—All organic substances contain C. One large and important class contain C and H alone; many consist of the three elements C, H, and O; others consist of C, H, O, and N; while a few contain also S and P. In the laboratory, organic compounds have been made which contain many other elements; but those given above exhaust the variety of elementary constitution in most natural organic substances.

The Number of Carbon Compounds greatly exceeds that of all the other substances combined, and is constantly increasing. The labor of modern chemists is largely devoted to the subject, and the field opens and broadens with every discovery. That such a vast number of different compounds of carbon with two or three other elements should exist, seems at first very puzzling. In inorganic chemistry, the number of compounds which any single element can form with *all* the others, is very limited. The explanation must be looked for, first of all, in a peculiarity of C.

We have already learned that C is quadrivalent, forming a compound with four atoms of H,— CH_4. If we use a dash to indicate each unit of valence, this formula may be written:

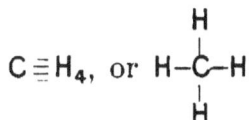

$$C \equiv H_4, \text{ or } H-\overset{\overset{\displaystyle H}{|}}{\underset{\underset{\displaystyle H}{|}}{C}}-H$$

Another compound of C and H is C_2H_6. The only way in which this can be expressed in detail, accounting for the quadrivalency of C in both atoms, is $H_3 \equiv C-C \equiv H_3$, or

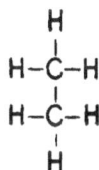

$$H-\overset{\overset{\displaystyle H}{|}}{\underset{\displaystyle |}{C}}-H \\ H-\overset{\displaystyle |}{\underset{\underset{\displaystyle H}{|}}{C}}-H$$

which shows the C atoms *linked together* by one dash or "bond" of valence, and thus capable of holding three, and only three, atoms of H apiece. In the same way, C_3H_8 equals $H_3 \equiv C-C=H_2-C \equiv H_3$; C_4H_{10} equals $C \equiv H_3 - C = H_2 - C = H_2 - C \equiv H_3$, etc.

In another series of *hydrocarbons*, a linkage between two atoms of C by double bonds occurs; thus, C_2H_4 (see p. 72), equals $H_2=C=C=H_2$; C_3H_6 equals $H_2=C=C-H-C\equiv H_3$, etc. In still another series, there is a linkage by three bonds; thus, C_2H_2 equals $H-C\equiv C-H$, etc.

It is through this facility with which C atoms link together in chains, that the number of different organic compounds is in part explained. When we consider further that one or more H atoms in each of these hydrocarbons may be replaced by O, N, or groups containing O, N, H, and C, such as OH, NO_2, CO, NH_2, CN, CH_3, etc., etc., the great variety of organic substances ceases to be so surprising.

Isomerism.—Isomeric compounds are those which consist of the same elements in the same proportion. Thus two compounds are known which have the same formula, $C_2H_4Cl_2$, and there are many similar cases. The difference between such compounds is believed to lie in a dissimilar grouping of the atoms about one another, as the same pieces upon a checker-board may be variously arranged; or as the letters p-l-e-a may also spell l-e-a-p, or p-e-a-l, or p-a-l-e; and this difference finds expression in the *constitutional formulas** of the substances. These formulas for the two compounds $C_2H_4Cl_2$, are:

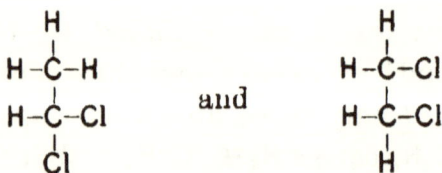

```
      H                        H
      |                        |
  H --C-- H                H --C-- Cl
      |          and           |
  H --C-- Cl               H --C-- Cl
      |                        |
      Cl                       H
```

Complexity of Organic Molecules.—While inorganic molecules consist of only a few atoms, and are therefore very simple in their construction, as:

* These formulas must not be understood to represent in any sense the relative *positions* of the atoms, but only the relations which they bear to one another, as shown by chemical reactions.

H_2O, CO_2; organic frequently contain a large number, and are extremely complex, as: Sugar $= C_{12}H_{22}O_{11}$, having 45 atoms in a molecule; stearin $= C_{57}H_{110}O_6$, having 173 atoms; albumin $= C_{72}H_{110}N_{18}SO_{22}$, having 222 atoms, and perhaps even more.

----•----

THE PARAFFINES AND THEIR DERIVATIVES.

MARSH-GAS, CH_4 (page 71), is the first member of a series of hydrocarbons whose common difference is CH_2, and whose general formula may be written C_nH_{2n+2}. Paraffine (*parum*, little; *affinis*, affinity), so called because acids and bases have no effect upon it, is a mixture of hydrocarbons of this series, and gives it its name. All of the members of the paraffine series are characterized by their chemical indifference. In the table below are given the names and formulas of the lowest members of the series:

PARAFFINE HYDROCARBONS, C_nH_{2n+2}.

		BOILING POINT.
Methane (Marsh-gas)	CH_4	Gas.
Ethane	C_2H_6	Gas.
Propane.................	C_3H_8	Gas.
Butane	C_4H_{10}	$1°$
Pentane	C_5H_{12}	$37°$
Hexane	C_6H_{14}	$71.5°$
Heptane	C_7H_{16}	$98°$
Octane................. ...	C_8H_{18}	$124°$
Nonane	C_9H_{20}	$150.8°$
Etc.		Etc.

All the hydrocarbons in this list, as well as many more of this series, occur in petroleum, but it is almost impossible to isolate any one completely.

Petroleum (*petra*, a rock; *oleum*, oil) is probably the product of the distillation of organic matter beneath the surface of the earth. It is not always connected with coal, as it is often found outside the coal-measures, as in New York and Canada. The distillation must have taken place at a much greater depth than that at which the oil is now found, as it would naturally rise through the fissures of the rock and gather in the cavities above. Sometimes the oil has collected on the surface of subterranean pools of salt-water, so that after a time the oil is exhausted, and salt-water only is pumped up; or if the well strikes the lower part of the cavity, the water will first be pumped, and afterward the oil. The crude oil from the well is purified by distillation, and treatment with concentrated H_2SO_4 and alkalies. By the distillation it is divided into several parts (each a mixture of hydrocarbons), which are called cymogene, rhigoline, gasoline, naphtha, benzine, kerosene.*

* Kerosene contains C_9H_{20}—$C_{16}H_{34}$. Kerosene accidents generally rise from the presence of naphtha. This is a cheap, light, dangerous oil. Its vapor, however, is not explosive unless mixed with air. While a lamp, which contains adulterated kerosene, is burning quietly, there is no danger. The vapor rises from the oil, fills the empty space in the lamp, but being unmixed with air, can not explode. Let, however, a draught of cold air strike it, or carry it into a cold room—instantly the vapor will be condensed, the air will rush in, and a dangerous mixture be formed. Or when the light is extinguished at night the vapor will cool, air pass in, and a mixture be produced which will be ready to explode when the lamp is relighted. Properly purified, kerosene is no more explosive than water, and will even extinguish a flame applied to it at the ordinary temperature. Dr. Nichols

The portion which is heavier than kerosene and boils at a higher temperature, yields lubricating oil and paraffine.

Paraffine is a hard white, tasteless solid, melting at 44° C. It is used for making candles. It was discovered in 1830, as a product of beech-wood, but all of the paraffine of commerce is now obtained from petroleum.

Bitumen or Asphaltum is another natural product, consisting chiefly of hydrocarbons. It is found in many parts of the world, sometimes pure, sometimes associated with various minerals. On the island of Trinidad is a lake of bitumen one and a half miles in circumference. Near the shore it is hard and compact, except in hot weather, when it becomes sticky. At the center it is soft, and fresh bitumen boils up to the surface. Asphaltum is found in immense quantities in California and in Canada. The bitumen is used for the same purposes as pitch, which it closely resembles. It is a natural cement for laying stone or brick. It was used in building the walls of Babylon, for which purpose it was gathered from the fountain of Is, on the banks of the Euphrates. It was a prominent ingredient in the "Greek Fire," so much used by the nations of Eastern Europe in their naval wars, even as late as the fourteenth century. This consisted of bitumen, sul-

gives the following simple test: *Fill a bowl partly full of hot water. Insert a thermometer, and add cold water until the temperature is 110° F. Then pour into the bowl a spoonful of kerosene, and apply a lighted match. If it takes fire, the oil contains naphtha and is dangerous.* See also an article in *Popular Science Monthly* for February, 1884.

phur, and pitch, and was thrown through long, copper tubes, from hideous figures erected on the prow of the vessel. Bitumen is used in making the famous promenades of the boulevards in Paris.

Artificial Preparation of the Paraffines.—Methane, CH_4, may be made by leading vapor of carbon disulphide, CS_2, mixed with H_2S, over heated Cu:

$$CS_2 + 2H_2S + 8Cu = CH_4 + 4Cu_2S.$$

This reaction is very interesting, because it shows that it is possible to make this organic compound from the elements; for CS_2 is prepared by the action of S vapor on C (page 110), and H_2S can be produced by passing H through boiling S, or by burning S vapor in an atmosphere of H.

The interest in this reaction becomes still greater when we learn that the other members of the series can be made from CH_4. The way in which this may be done is the following: 1st. One atom of H in CH_4 is replaced by an atom of Cl, by the action of Cl on CH_4, giving the compound CH_3Cl. 2d. Two molecules of CH_3Cl are acted upon by Na:

$$2CH_3Cl + 2Na = 2NaCl + C_2H_6.$$

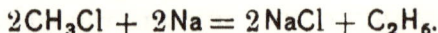

This gives the second member of the series, ethane, in which two C atoms are linked. Then by a similar process the higher members may be formed, thus:

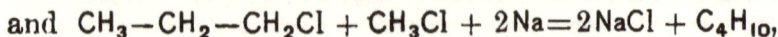

$$CH_3—CH_2Cl + CH_3Cl + 2Na = 2NaCl + C_3H_8;$$

and $$CH_3—CH_2—CH_2Cl + CH_3Cl + 2Na = 2NaCl + C_4H_{10},$$

or, $$2CH_3—CH_2Cl + 2Na = 2NaCl + C_4H_{10}.$$

THE OXYGEN COMPOUNDS OF THE PARAFFINES.

THE ALCOHOLS.

THE alcohols are compounds which may be regarded as hydrocarbons, in which the univalent group OH, called *hydroxyl*, has replaced H. It is convenient to consider the alcohols as hydroxides of certain groups or *radicals*. Thus the alcohol derived from methane, CH_3OH, is regarded as the hydroxide of the radical CH_3, which is called *methyl*; the alcohol from ethane, C_2H_5OH, as the hydroxide of C_2H_5, *ethyl*, and so on; and the alcohols themselves are commonly called methyl alcohol, ethyl alcohol, etc. These organic radicals can not be isolated. They are analogous to the inorganic group or radical NH_4 (page 136), and the analogy of the alcohols to metallic hydroxides is readily seen. The alcohols, as well as the other derivatives of the paraffines, can be artificially prepared.

Methyl Alcohol, CH_3OH, is obtained as one of the products of the destructive distillation of wood.* It is a light, volatile liquid, which closely resembles

---

* When hard wood, as beech or oak, is heated to a high temperature, with no O present, or an imperfect supply, it is decomposed; the charcoal remains, while a large number of products is formed, among which are H, CO, CO_2, H_2O, CH_4, methyl alcohol, acetic or pyroligneous acid, creosote, paraffine, tar, etc.

Creosote (flesh-preserver) is a colorless, poisonous liquid, with a flavor of burnt wood. It has powerful antiseptic properties. It imparts to smoke a characteristic odor, renders it irritating to the eyes, and also gives to it the power which it possesses of curing hams, beef, etc. Much of that which is sold as creosote is carbolic acid. (See page 223.)

ordinary alcohol in all its properties. It is used in the manufacture of aniline dyes, in making varnishes, and in spirit-lamps.

Ethyl Alcohol, $CH_3—CH_2OH$, *ordinary alcohol*, is the best known and most important of the alcohols. It is formed by the fermentation of saccharine substances, and prepared by distillation of the solution thus produced. In this way an alcohol of about 93 per cent. can be obtained. In order to get "absolute" alcohol, this product must be mixed with quick-lime, which retains the water, and again distilled. Pure alcohol boils at 78° C., and has recently been frozen at a temperature of —130.5° C.

When it is exposed to the air the spirit evaporates, while moisture is absorbed from the atmosphere.* It burns without smoke and with great heat, owing to the abundance of H and deficiency of C, and is therefore of much value in the arts. It is also of incalculable importance as a solvent of many substances —roots, resins, fragrant oils, etc.

Effects of Alcohol.—When pure it is a deadly poison. When diluted, as in the ordinary liquors, it is stimulative and intoxicating. Its influence is on the brain and nervous system;—deadening the natural affections, dulling the intellectual operations and moral instincts; seeming to pervert and destroy all that is pure and holy in man, while it robs him of his highest attribute—reason. (See "Physiology," p. 150.)

---

* The chemist discovers this when he neglects to put the extinguisher on his alcohol lamp, and finds that he can not relight it without moistening the wick with fresh alcohol.

FERMENTATION.

IF a pure solution of grape-sugar be exposed to the air it will undergo no change; but if there be added a little ferment,* or any albuminous substance (*i. e.*, one containing N), in a decomposing state, it will immediately commence breaking up into new compounds. The fermentation is due to the presence of small organized bodies, which find materials for their sustenance in the solution, and in their growth overcome the equilibrium of the chemical forces, causing the large molecules to drop into smaller ones. There are different kinds of ferments, which cause different kinds of fermentation.

1st. **Alcoholic Fermentation.**—In this, *the grape-sugar is resolved into alcohol and carbon dioxide.* The former remains in the liquid, while the latter escapes in little bubbles of gas. The reaction may be represented thus : $C_6H_{12}O_6 = 2C_2H_6O + 2CO_2$.

2d. **Acetic Fermentation.†**—This often succeeds

---

* "In many cases, spontaneous fermentation sets in without the apparent addition of any ferment: thus wine, beer, milk, etc., when allowed simply to stand exposed to the air, become sour, or otherwise decompose. These changes are, however, not effected without the presence of vegetable or animal life, and are true fermentations; the *sporules*, or seeds of these living bodies, always float about in the air, and on dropping into the liquid begin to propagate themselves, and in the act of growing evolve the products of the fermentation. If the above liquids be left only in contact with air which has been passed through a red-hot platinum tube, and thus the living sporules destroyed ; or if the air be simply filtered by passing through cotton wool, and the sporules prevented from coming into the liquid, it is found that these fermentable liquids may be preserved for any length of time without undergoing the slightest change."—ROSCOE.

† There are also other forms of fermentation, as the lactic, yielding lactic acid—the acid of sour milk; butyric, yielding butyric acid, etc.

the first immediately, if not checked, and the alcohol is broken up into acetic acid and water ; thus, $C_2H_6O +$ O_2 (from the air) $= C_2H_4O_2 + H_2O.$

Yeast is the ferment which causes alcoholic fermentation. It consists of microscopic plants (*Saccharomyces cerevisiæ*), which increase by the formation of multitudes of tiny cells not more than $\frac{1}{2400}$ of an inch in diameter. In the brewing of beer they grow in great abundance, making common brewer's yeast.*

Malt.—In making malt, the barley is thoroughly soaked in water, and then spread on the floor of a dark room, to heat and sprout. Here a curious change ensues, identical with that which takes place -in every planted seed. Each one contains starch and a nitrogenous substance called *gluten.* The tiny plant not being able to support itself in the beginning, has here a little patrimony with which to start in life ; but, as the starch is insoluble in the sap, it must first be changed to a soluble form. We see, therefore, the need of a ferment ; but it would not answer to store up in the seed an active ferment, as that might cause a change before the plant was ready to grow, and thus the plant's capital be wasted. The gluten acts, therefore, as a *latent* ferment. As soon as the seed is planted it absorbs moisture from the ground, is turned into *diastase—* an *active* ferment †—the starch is converted into

---

* The yeast-cakes of the kitchen are formed by exposing moistened Indian meal, containing a ferment, to a moderate temperature, until the gluten or albuminous matter of the cake has undergone this alcoholic fermentation. They are then laid aside for use.

† "Malt does not contain more than $\frac{1}{500}$ of its weight of diastase ; one

dextrin and sugar, dissolved, and immediately applied to the uses of the growing plant. This change takes place in the *malting-room*. The barley sprouts, and a part of its starch is turned to sugar, so as to give it a sweetish taste. If this germination were allowed to proceed, the little barley sprout would turn the sugar into woody fiber. To prevent this, the grain is heated in a kiln until the germ is destroyed. Barley in this condition is called *malt*, and is then transported to the breweries.

Brewing Beer.—The malt is crushed and digested in water, to convert the remaining starch into dextrin and sugar. Hops and yeast are added, and fermentation immediately commences. Bubbles of gas rise to the top with a low hissing sound, yeast gathers in a foamy cream that comes to the surface of the tub, while the alcohol gradually accumulates in the liquid. The beer is now drawn off into tight casks, where it undergoes a second fermentation ; the flavor ripens, and the CO_2 collecting, gives to the liquor, when drawn, its sparkling, foamy appearance.

Lager Beer (*Lagern*, to lie) is so called because it is allowed to lie for months in a cool cellar, where it ripens very gradually. It is also fermented much more slowly and perfectly than ale or porter.

Wine is made from the juice of the grape. The juice, or *must*, as it is called, is placed in vats, in the cellar, where the low temperature produces a slow fermentation. When all the sugar is converted into

part of this substance being sufficient to change 2,000 parts of starch into dextrin and sugar."—PERSOZ AND PAYEN.

alcohol and CO_2, a dry wine remains; when the fermentation is checked, a sweet wine is the result; and when bottled while the change is still going on, a brisk, effervescing wine, like champagne, is formed. The flavor or "bouquet" of wine is due to the slow formation of a fragrant and aromatic ether.* (See p. 206). The tartaric acid of the grape gradually separates and collects on the sides and bottoms of the casks in an incrustation—tartar, an impure acid potassium tartrate, from which tartaric acid and cream of tartar are made.

Alcohol in Beer, Wine, etc.—Alcohol is the intoxicating principle of all varieties of liquors, ale, beer, wine, cider, and the domestic wines. Ale and porter contain from 6 to 8 per cent. of alcohol; wine varies from 7 per cent. in the light claret to 17 per cent. in the strong Port and Madeira; brandy and whiskey have from 40 to 50 per cent.

Ardent Spirits.—When any fermented liquor is distilled, the alcohol passes over, together with water and some fragrant substances which are condensed. In this way brandy is made from wine; rum, from fermented molasses or cane-juice; whiskey, from fermented corn, rye, or potatoes; and gin, from fermented barley and rye, afterward redistilled with juniper-berries. The accompanying cut represents an apparatus used for this distillation. A is the boiler, B the dome, C a tube passing into S, the condenser, where it is twisted into a spiral form called the

---

* Œnanthic ether, a liquid with a powerful odor, which causes the peculiar smell of grape-wine.

worm, in which the vapor from the boiler is condensed, and drops out at D. (See "Physics," page 188.)

FIG. 69.

A Still.

Amyl Alcohol, $C_5H_{11}OH$, is the chief constituent of "fusel oil," found in whiskey distilled from potatoes. It is often present in common alcohol, giving a slightly unpleasant odor when it evaporates from the hand. It is extremely poisonous, and as it is often contained in liquors, must greatly increase their destructive and intoxicating properties.

Besides these alcohols which have been described, and others which, like them, contain but one hydroxyl group, there are alcohols which have two, three, and more OH groups. Among these the most important one is *glycerin*, $C_3H_5(OH)_3$, which is a constituent of fats, and will be treated later (page 206).

THE ALDEHYDES AND ACIDS.

WHEN alcohols are oxidized they are converted into acids. By taking proper precautions, intermediate products, called aldehydes (*alcohol dehydrogenatum*), can be obtained from alcohols like those we have considered. The two stages of the oxidation are shown by the following equations:

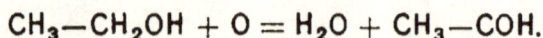

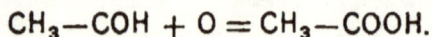

$$CH_3-CH_2OH + O = H_2O + CH_3-COH.$$

$$CH_3-COH + O = CH_3-COOH.$$

The group $-C=O.H$ is characteristic of the aldehydes; the group $-C=O.OH$, "the carboxyl group" of the organic acids.

Ethyl Aldehyde, CH_3-COH, is a colorless liquid boiling at 21° C. It is readily oxidized to acetic acid or reduced to ethyl alcohol. It has a peculiar and characteristic odor, which may be obtained by holding a red-hot coil of Pt wire in a goblet containing a few drops of alcohol.

Formic Acid, CH_2O_2 or $HC=O.OH$, occurs in red ants (*formica rufa*) and stinging nettles, and can be obtained from them by distillation with water. It is best prepared by heating oxalic acid with glycerin.* It is a fiery, pungent liquid, which blisters the skin. It is a monobasic acid (page 112), the H of the carboxyl group being alone replaceable by metals, and yields salts called *formates*, *E. g.* $H-C=O.OK$, potassium formate.

* It can also be made by the oxidation of methyl alcohol, CH_3OH.

Acetic Acid, $C_2H_4O_2$, or $CH_3-C=O.OH$ (*acetum,* vinegar), forms from two to four per cent. of common vinegar, whence its name. The strongest acetic acid is known as the *glacial,* since it crystallizes into

FIG. 70.

an ice-like solid at 17° C. It has an aromatic taste and pungent odor, and, after a time, blisters the skin.

Making Vinegar.

Preparation.—Vinegar is made on a large scale by allowing a weak alcohol to trickle slowly through a cask filled with beech shavings, which have been soaked in vinegar. As the alcohol passes down, it is oxidized, and after two or three repetitions cf the process, it becomes entirely converted into vinegar.*

Cider Vinegar.—By the alcoholic fermentation, sweet cider becomes "*old cider.*" By exposure to the air the alcohol passes on to the second stage, and the acetic acid formed produces the sour taste of the vinegar.

Pyroligneous Acid, wood-vinegar (see p. 192, note), is crude acetic acid. It is used in making acetates, from which the pure acid is obtained by the action of a stronger acid, as H_2SO_4.

Properties.—Acetic acid is monobasic. One of its most important salts is "sugar of lead" ($CH_3-C=O.O)_2Pb$ (see p. 164). Vinegars of commerce are

---

* The oxidation of alcohol into vinegar is due to a microscopic organism (*mycoderma aceti*), which is called "mother of vinegar." It conveys O from the air to the alcohol. In the "quick-vinegar process" described, the organism is deposited on the shavings from the vinegar with which they are soaked.

often sharpened by the addition of H_2SO_4 and pungent spices.*

Its sole use as a food is as a condiment. It allays thirst, and was anciently carried by the Roman soldiers in a little flask for that purpose. Sugar added to vinegar quickly passes to the second stage of fermentation, and increases its strength. Indeed, vinegar is sometimes made entirely from sweetened water and tea-leaves, which act as a ferment. It prevents the decomposition of both animal and vegetable substances, and is hence used for preserving them.

Preserves frequently "work," as it is called, and then sour. The bubbles of gas which rise to the surface indicate the alcoholic fermentation. If neglected, this soon passes to the acetic stage. It may be checked by scalding, which destroys the ferment.

——————— ———————

Of the other acids of this series containing one carboxyl group, the only ones of especial interest are *Palmitic acid* $C_{15}H_{31}CO.OH$, and *Stearic acid* $C_{17}H_{35}CO.OH$, which, in combination with glycerin, form the common solid fats. Another related acid, the *Oleïc*, is a constituent of the liquid fats (see p. 206). The so-called "stearin" candles are made of a mixture of palmitic and stearic acids.

* We can easily detect these by evaporating a half-gill in a saucer, placed over hot water. As it boils down, add a little sugar, taking care not to allow it to burn. If the liquid turns black, it is proof of the presence of H_2SO_4. As the last evaporates, the odor of cayenne pepper, etc. (if there be any), can be readily distinguished. In England, commercial vinegar is permitted by law to have one part in a thousand of oil of vitriol, as this keeps it from molding.

A few important acids, which may be regarded as derivatives of the paraffine series, and which contain two or more carboxyl groups, are treated below.

Oxalic Acid, $C_2H_2O_4(C=O.OH)_2$, is familiar in the sour taste of rhubarb, sorrel, etc. In these plants the acid is combined with K and Ca. It may be prepared by the action of HNO_3 on sugar.* It is a potent poison. The antidote is powdered magnesia, or chalk, stirred in H_2O. It is a test of lime, forming a delicate white precipitate of calcium oxalate. A solution of oxalic acid is much used to remove ink stains, and is often sold for this purpose under the deceptive name of "salts of lemon." The acid unites with the Fe of the ink, and the iron oxalate thus made is soluble in H_2O. It should be washed out immediately, as it will corrode the cloth.

Malic Acid $\left(C_4H_6O_5, \begin{array}{l} CH(OH)CO.OH \\ | \\ CH_2CO.OH \end{array}\right)$, occurs abundantly in most acid fruits, particularly in unripe apples, whence its name from *malum*, an apple. *Citric acid*, $C_3H_4(OH)(CO.OH)_3$ (*citrus*, a lemon), the acid of the lemon, lime, etc., is often found associated with it, as in the gooseberry, raspberry, and strawberry. Citric acid is used in medicine as magnesium citrate.

Tartaric Acid $\left(C_4H_6O_6, \begin{array}{l} CH(OH)CO.OH \\ | \\ CH(OH)CO.OH \end{array}\right)$, exists in

---

* Oxalic acid is made on a large scale from sawdust, soda, and caustic potash. The woody fiber is resolved into oxalic acid, which combines with the bases, forming sodium and potassium oxalates. From these the acid is readily obtained. Sawdust will yield more than half its weight of crystals of this salt.

many fruits, principally in the grape, combined with
K as acid potassium tartrate ("bitartrate of potash").
This settles during the fermentation of wine (see p.
198), and when purified is called *cream of tartar*,
from which the acid is prepared. It forms transparent crystals of a pleasant acid taste, which are
permanent in the air. Its aqueous solution gradually
becomes moldy, and spoils. *Tartar emetic* is an
antimony potassium tartrate. *Rochelle salt* is a sodium potassium tartrate; it is commonly used in
medicine in the form of *Seidlitz powders*. These are
contained in a blue and a white paper. The former
holds 120 grains of Rochelle salt, and 40 grains of
bicarbonate of soda; the latter, 35 grains of tartaric
acid. They are dissolved in separate goblets. The
one containing the acid is emptied into the other,
when the CO_2 is set free, producing a violent effervescence, and disguising the taste of the medicine.

THE ETHERS AND ETHEREAL SALTS.

THE ethers are oxides of the organic radicals.
They are thus analogous to the metallic oxides, just
as the alcohols are to the metallic hydroxides, and
are related to the alcohols in the same way that K_2O
is related to KOH. Each of the alcohols has its corresponding ether. Thus we find:

Methyl ether $(CH_3)_2O$, corresponding to methyl alcohol, CH_3OH.
Ethyl ether $(C_2H_5)_2O$, corresponding to ethyl alcohol, C_2H_5OH.
Propyl ether $(C_3H_7)_2O$, corresponding to propyl alcohol, C_3H_7OH.
Etc., etc.

Such ethers as these, in which two radicals of the same kind are united with O, are called *simple ethers*. It is easy to make ethers which contain different radicals, as, for instance, $CH_3-O-C_2H_5$; and in such cases they are called *mixed ethers*.

Ethyl Ether $(C_2H_5)_2O$, *Common or "Sulphuric" Ether*,* is made by heating a mixture of common alcohol and H_2SO_4. It is a colorless, very volatile liquid of a peculiar odor. It is much lighter than water, and somewhat soluble in it. Ether boils at 35° C. Its vapor is thirty-seven times heavier than hydrogen, and can be poured like CO_2. It is very inflammable, and its vapor, with air, forms an explosive mixture.† Ether is a valuable anæsthetic, and is extensively used in surgery.

Ethereal Salts are salts in which an organic radical occupies the place of the metal in an ordinary salt. Thus, in making ether from alcohol by the action of H_2SO_4, there is first formed $\overset{C_2H_5}{\underset{H}{}}SO_4$, in which the radical C_2H_5 has replaced a H atom of the acid; it is analogous to $KHSO_4$, which is formed by the action of H_2SO_4 on KOH:

$$KOH + H_2SO_4 = H_2O + KHSO_4.$$
$$C_2H_5OH + H_2SO_4 = H_2O + (C_2H_5)HSO_4.$$

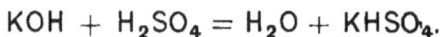

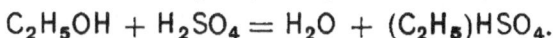

As other examples of ethereal salts, or *compound ethers*, as they are often termed, we have: Ethyl sulphate $(C_2H_5)_2SO_4$ (analogous to K_2SO_4); ethyl ni-

* This name was given to it because H_2SO_4 is used in its manufacture, and not because it contains any S.
† Never use ether in the neighborhood of flames.

trate, $C_2H_5NO_3$; ethyl chloride, C_2H_5Cl; ethyl acetate, $CH_3—CO.OC_2H_5$.

A number of ethereal salts are extensively sold as flavoring extracts for the use of confectioners and cooks. The essence of jargonelle pear is an alcoholic solution of amyl acetate; apple oil, of amyl valerianate; pine-apple, of ethyl butyrate.

To this same class of compounds belong also the natural fats, most of which are mixtures of the ethereal salts which glycerin forms with palmitic, stearic, and oleïc acids, and which are called *palmitin*, *stearin*, and *oleïn*. Butter consists of the glycerin salts of seven acids, all derivatives of the paraffine series.

The Fats and Glycerin. — *Palmitin* and *Stearin* are solids, while oleïn is liquid. Therefore, the larger the proportion of oleïn which a fat contains, the lower its melting point, and the softer it is. Fats especially rich in palmitin are human fat and palm oil; in stearin, mutton tallow, beef tallow, and lard; in oleïn, sperm-oil, and codliver oil.

Glycerin is an alcohol containing three OH groups, $C_3H(OH)_3$, and its ethereal salts, with the fatty acids, are:

$(C_{15}H_{31}CO.O)_3C_3H_5$, glycerin tripalmitate or palmitin,
$(C_{17}H_{35}CO.O)_3C_3H_5$, glycerin tristearate or stearin,
$(C_{18}H_{33}CO.O)_3C_3H_5$, glycerin trioleate or oleïn.

Glycerin is obtained from the natural fats by heating them with lime or lead oxide, or by the action of superheated steam. In the former cases, the calcium or lead salt of the acid is formed, and the glycerin set free; in the latter case, the fat is decomposed at once into acid and glycerin. The acids, when cool, are subjected to great pressure; the oleïc flows out, leaving the stearic and palmitic acids as a milk-white, odorless, tasteless solid, which is commonly called *stearin,* and extensively used in the manufacture of *stearin* or *adamantine candles.**

Glycerin is an odorless, transparent syrup. It is soluble in H_2O and alcohol. On account of its healing properties, its use is common in dressing wounds, insect bites, chapped hands, etc.

By the action of HNO_3 and H_2SO_4, glycerin is converted into *nitro-glycerin* ($C_3H_5(NO_3)_3$, glycerin nitrate, a compound ether), an oil that often explodes with fearful violence by a slight concussion. *Dynamite,* used in blasting, is powdered silex, or infusorial earth ("Geology," p. 48), saturated with nitro-glycerin.

Soap.—If sweet-oil and H_2O be placed in a test-tube and shaken, they will mix, but not unite; for on standing, the oil will rise to the top. Add, however, caustic potash or a little "lye" (see p. 129), when, on heating, a clear, soapy solution will be

* *Wax candles* are manufactured by the following process: A large number of cotton wicks are hung upon a revolving frame with projecting arms. The wicks are fitted at the ends with metal tags to keep the wax from covering that part. As the machine slowly turns, a man, standing ready with a vessel of melted wax, carefully pours a little upon each wick in succession. This process is repeated until the candles reach the desired size. They are then rolled on a smooth stone slab, the tops cut by conical tubes, and the bottoms trimmed, when they are ready for use. The large tapers burned in Catholic cathedrals are made by placing the wick on a sheet of wax, rolling it up till the right thickness is reached, when the candle is trimmed and polished as before. *Spermaceti candles* are run from the white, crystalline, solid fat which is found with sperm-oil in the head of the sperm-whale.

formed. The K of the alkali has combined with the
oleïc and palmitic acids of the oil, making two new
salts — potassium oleate and potassium palmitate;
while the expelled glycerin remains floating in the
liquid.

The manufacture of soap is based on this principle.* A variation in the alkaline base and the fat
or oil used, produces the different kinds of soap.
Potash, on account of its affinity for H_2O, forms *soft-
soap.* Soda is not deliquescent,† and hence makes
hard-soap.‡ Lard forms a softer soap than tallow.
Castile soap is made from olive-oil and soda. Its
mottled appearance is due to the coloring matter
which is stirred through it while it is yet soft.
Home-made soap is prepared by boiling "lye" and
"grease."§ As the latter contains such a variety of
fatty substances, the soap generally consists of the
three salts—potassium oleate, palmitate, and stearate.
Yellow soap contains some resin in place of fat.
Cocoanut-oil makes a soap which will dissolve in salt
water, as it contains an excess of alkali. *Soap-balls*
are made by dissolving soap in a very little water,

---

* Saponification (*sapo*, soap; *facere*, to make) is the process of separating
the fatty acids and glycerin, and is so named even when no soap is formed.
One method is as follows: Tallow or lard is boiled with lime, and thus
made into a calcium soap. This is decomposed by H_2SO_4, forming calcium sulphate, which, being insoluble, sinks to the bottom, leaving the
three acids of the fat floating upon the surface.

† A deliquescent substance is one that dissolves in H_2O, which it absorbs from the air.

‡ Soap is frequently adulterated with gypsum, lime, pipe-clay, or sodium
silicate. These may be detected by dissolving a piece of the soap in distilled water or alcohol, and noticing if there be any precipitate.

§ The heat hastens the chemical change, which takes place more slowly
in making what is known as "cold soap."

and then working it with starch to a proper consistency to be shaped into balls. *White toilet-soaps* are made from lard and soda. The *curdling of soap* in hard water is caused by the formation of a calcium or a magnesium soap, which is insoluble in H_2O, and floats on the top as a greasy scum.*

The Cleansing Qualities of Soap.—There exudes constantly from the pores of the skin an oily perspiration, and this, catching the floating dust, dries into a film which will not dissolve in H_2O. The alkali of the soap combines with this oily substance, and makes a soluble soap. In addition, the alkali also dissolves the cuticle of the skin, and thus produces the "soapy feeling," as we term it, when we handle soap.

HALOGEN DERIVATIVES OF THE PARAFFINES.

THE halogens can be caused to replace one or more atoms of H in the paraffines, and many of their derivatives. A few only of the compounds thus formed are of practical importance.

Chloroform, $CHCl_3$, *tri-chlor-methane*, is prepared by distilling alcohol, C_2H_5OH, with chloride of lime. It is a colorless, heavy liquid of a sweetish taste and ethereal odor. It is scarcely soluble in water, and when shaken with it quickly settles out at the bottom. It should evaporate without any unpleasant

* A soap made from lard, in water containing calcium carbonate, would undergo the following reaction : Potassium oleate + calcium carbonate = calcium oleate + potassium carbonate.

odor or residue. It boils at 62° C. It is very useful as an anæsthetic, and is employed as a solvent of I, P, S, caoutchouc, fatty and resinous bodies.

Iodoform, CHI₃, made by bringing together sodium carbonate, alcohol and iodine, is a solid. It forms small yellow crystals, which melt at 119° C. It is used in surgery.

Chloral, CCl₃—CHO, *tri-chlor-aldehyde*, is formed by passing Cl through absolute alcohol. It is an oily liquid which combines with H₂O, making *Chloral Hydrate*, a white, crystalline substance, much used to induce sleep. Taken in proper quantities, it is entirely safe, and is exceedingly pleasant in its influence.

THE CARBOHYDRATES.

STARCH, WOODY FIBER, AND SUGAR.

1. STARCH ($C_6H_{10}O_5$).

Source.—Plants accumulate it, 1, in their roots, as the carrot, the turnip, etc.; 2, in subterranean stems, as the potato, of which it forms about 20 per cent.; 3, in the base of their leaves, as the onion; 4, in the seed, as corn, of which it forms about 65 per cent., the bean, the pea, etc. In all these it is stored up for the future growth of the plant. It is kept in its starch form (lest it dissolve in the first rain), and then turned to sugar only when the

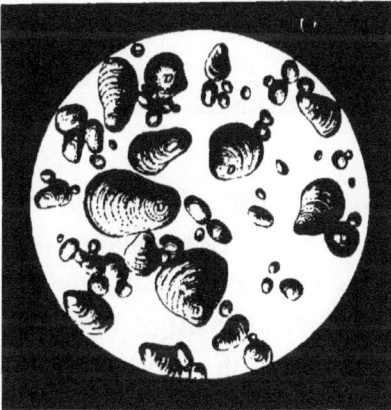

FIG. 71.

Potato Starch.

plant needs it in growing. (See p. 196.) Under the microscope, each vegetable is found to have its peculiar form of starch granule, so that in this way any adulteration is easily detected.*

* "The structure of the grains of starch is very beautifully displayed by placing some of them in contact with a drop of concentrated solution of zinc chloride (tinged with a little free iodine) on the field of the micro-

Preparation.—Starch is made from wheat, corn, potatoes, etc. The process
is essentially the same in
all. The potato, for ex-
ample, is ground to a
pulp, and then washed
with cold water. The
starch settles from this
milky mass as a fine,
white precipitate.

FIG. 72.

Wheat Starch.

Properties.—Starch is
insoluble in cold water;
in hot, it absorbs H_2O,
swells, and the granules
burst, forming a jelly-like liquid, used for *starching.*

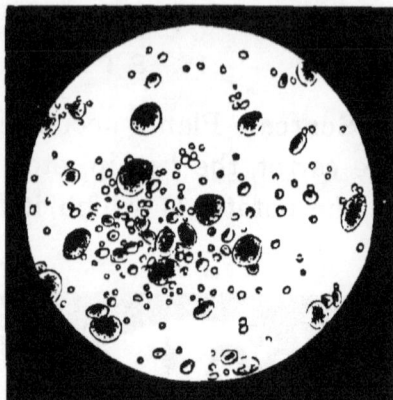

FIG. 73.

Bursting of Starch Granule.

The swelling of rice, beans, etc.,
when cooked, is owing to this
property. By heating to 400°
when dry, starch undergoes a
peculiar change into a substance
known as dextrin,* used as a
mucilage on envelopes and adhe-
sive stamps, for making "fig-
paste," and stiffening chintzes. The test of starch is
I, which forms in solution the blue iodide of starch.

scope. No change takes place in the granules until a little water is added.
They then become of a deep blue color, and gradually expand; at first, a
frill-like plaited margin is developed around the globules; by degrees this
opens out; the plaits upon the globule may then be seen slowly unfolding,
and may be traced in many cases into the wrinkles of the frill; ultimately
the granules swell up to twenty or thirty times their original bulk, and
present the appearance of a flaccid sac."—BUSK.

* Dextrin is isomeric with starch, but is not discolored by I.

Sago is the starch from the pith of the palm-tree; tapioca and arrow-root are made from the roots of South American marshy plants.*

Gum is found in the juices of nearly all plants, and frequently exudes, as in the peach, plum, and cherry. It is soluble in water, but not in alcohol. *Gum arabic*, which flows in transparent tears from an acacia tree, is the purest form.† *Mucilage*, which occurs in gum tragacanth, linseed, quince-seed, etc., is a modification of gum, and is insoluble in H_2O. It forms with it, however, a gelatinous liquid, which is exceedingly useful.

Vegetable Jelly.—"A variety of gum called *pectose* exists in nearly all fruits and vegetables. It gives to them their hardness while green."—FREMY. In the process of ripening, or by heat, acids, etc., it is turned into *pectin*. We find this abundant in the thick juice which exudes from an apple while baking. In the making of jellies, pectose is converted into a mixture of pectosic and pectic acids.

--------•●•--------

2. CELLULOSE ($C_6H_{10}O_5$).

Sources.—If a thin slice of wood be examined under the microscope, it will be seen to consist of a fibrous substance incrusted and compacted with

---

* Very many of the farinaceous preparations sold for the sick and invalid, under high-sounding names, are simply wheat or corn starch.

† It is a soluble salt, being composed of arabic acid ($C_{12}H_{22}O_{11}$, *Gélis*), combined with K and Ca.

woody matter. The former is called *cellulin* ($C_6H_{10}O_5$).* It composes the cells of all plants, giving them strength and firmness, and is found even in delicate fruits, holding their luscious juices. It occurs in various modifications, in wood, nut-shells, and fruit-stones. In the heart of a tree, its cells are hard and dense; in the outer part, they are soft and porous; in elder-pith and cork, light and spongy; in flax and cotton, long, pliable, and fibrous; in the bran of wheat and corn, digestible.

Secretion.—All vegetation consists of these simple cells. They seem alike to the eye, yet they have a very diverse power of secretion. The cell of the sugar-maple converts its sap into sugar; the milk-weed, into a milky juice; the caoutchouc, into rubber; the rhubarb-plant, into oxalic acid; and the rose-petal, into the most delicate of perfumes.

Cells are always true to themselves. There seems to be a law of God stamped on each one, so that when we cut a tiny bud from one tree and graft it into another, it remains consistent with itself. It develops into a limb, and years pass by. The few single cells become a myriad, yet they have not changed. The sap flows upward in the tree; but at a certain point—a hidden threshold which no human eye can discern—it comes under a new and strange influence. Here it is transformed, and produces fruit and flowers, in accordance with another and different growth. Somehow, quince-juice is made into pears, locust-

---

* It is probable that the molecule of woody fiber is some multiple of this formula, as $C_{18}H_{30}O_{15}$.

juice blooms out into fragrant acacias, and sweet and sour apples hang upon the same limb.

Uses. — These are wonderfully various. Woody fiber is woven into cloth, built into houses, twisted into rope, twine, and thread, made into paper, cut into fuel, carved into furniture. We eat it, wear it, walk on it, write on it, sit on it, print on it, pack our clothes in it, sleep in it, ride in it, and burn it.

Paper is made from cotton, linen, straw, or any substance containing cellular tissue. The finest writing-paper is manufactured from linen rags. These are first "shredded" upon scythe-blades—*i. e.*, the seams are ripped open, buttons cut off, and the dust shaken out. 2d. They are steamed in a solution of chloride of lime for ten or twelve hours, until they are thoroughly bleached. 3d. They are received by a machine that alternately lacerates them by a cylinder set with razor-like blades, and washes them with pure, cold water for six hours, until they are reduced to a mass resembling rice and milk. 4th. This pulp receives a delicate blue tint from *smalt*.* 5th. It is diluted with H_2O nearly to the consistency of milk, and strained to remove the waxed ends and knots of thread that cause the little lumps which catch our pen when we write rapidly on poor paper. 6th. It flows over an endless belt of wire-gauze, about thirty feet in length, through which the water steadily drips from the pulp, as it slowly passes along, gaining consistency and firmness. 7th. It comes to a

* Powdered glass colored with oxide of cobalt.

part of the belt called the "dandy-roll," consisting of a cylinder, on the surface of which are wires arranged in parallel rows, or fancy letters, which print upon the moist paper a design — constituting what is termed "laid," or "wire-woven," paper. 8th. The paper, very soft and moist as yet, passes between rollers that squeeze out the water; then between others which are hot, and dry it. 9th. It comes to a vat of sizing, composed of glue and alum, into which it plunges, and at the opposite side emerges only to go between other rollers, that press and dry it, at the end of which it passes under a cylinder set with knives, that clip the roll into sheets of any desired size.

Paper Parchment is prepared by plunging unsized paper for a few seconds in H_2SO_4, of a specified strength, then washing off the acid. This, in some unknown way, changes its appearance and character, so that it resembles parchment, while its toughness is five times that of the paper from which it was made.

Linen is made from the inner bark of flax. The plant is first *pulled* from the ground, to preserve the entire length of the stalk; next "rotted," by exposure to air and moisture, when the decayed outer bark is removed by "breaking"; then, by "hatcheling," the long, fine fibers are divided into shreds, and laid parallel, while the tangled ones are separated as "tow." It is then bleached on the grass, which renders the gray coloring-matter soluble by boiling in lye. The whitened flax is lastly woven into cloth.

Cotton consists of the beautiful hollow, white hairs arranged around the seed of the cotton-plant. As it is always pure and white—except Nankin cotton, which is yellow—it would require no bleaching did it not become soiled in the process of spinning, etc.

Gun-Cotton.—*Pyroxylin* (*pur*, fire, and *xulon*, wood) is prepared by dipping cellular tissue—cotton, sawdust, printing-paper, etc.—in a mixture of HNO_3 and H_2SO_4 of a certain specific gravity. It is then carefully washed and dried. It is not materially changed in appearance, but a part of its H has been replaced by NO_2, and it has become very inflammable. It burns very rapidly, and, unlike gunpowder, leaves no residue. It explodes on percussion with a force which is far greater than that of gunpowder. A mixture of gun-cotton and camphor is widely used under the name of *celluloid*.

Collodion is a solution of gun-cotton in ether and alcohol. It forms a syrupy liquid, which is much used by photographers.

3. SUGAR.

Cane-Sugar ($C_{12}H_{22}O_{11}$),* *Sucrose*, is obtained from the sap of the sugar-maple, and the juice of the sugar-cane, sorghum, and beet. In making it from

* A very brilliant experiment showing the presence of C in $C_{12}H_{22}O_{11}$ is obtained by putting on a clean, white plate, a mixture of finely-pulverized white sugar and $KClO_3$. Upon adding a few drops of H_2SO_4, a vivid combustion will ensue. By mixing with the sugar a few iron and steel filings, and performing the experiment in a dark room, or out-of-doors at night, fiery rosettes will flash through a rose-colored flame, and produce a fine effect.

the sugar-cane, the canes are crushed between iron cylinders, to express the juice. A little lime is added to neutralize the acids, which would prevent complete crystallization of the sugar, and to remove certain substances which would cause fermentation, and it is then evaporated to a thick jelly, and set aside to cool. The sugar crystallizes readily, forming *brown sugar*, which is put in perforated casks to drain. The drainings, or "mother-liquor," constitute molasses.

Refining.—Brown sugar is dissolved in H_2O, filtered through twilled cotton to remove the coarse impurities, and then through a deep layer of animal charcoal. The colorless solution is next evaporated in vacuum pans, from which the air is exhausted, so that the sugar boils at so low a temperature as to avoid all danger of burning. When sufficiently concentrated, the liquid is removed and set aside to crystallize. If the mass of crystals is dried in molds, it forms *loaf-sugar;* if in centrifugal machines, *granulated sugar.** The drainings constitute "syrup," "sugar-house molasses," etc.

Confectionery.—Terra alba (white earth) is imported from Ireland for use in lozenges, drops, etc.†

* This apparatus consists of a cylindrical drum mounted upon a vertical axis, to which a rapid rotary movement can be given. The outer side of this drum is made of a stout but closely-woven net-work. The drum is inclosed in a large, fixed, cylindrical vessel, capable of holding the liquid which may pass out through the net-work. A charge of sugar is placed in the inner drum, which is then made to revolve rapidly. The syrup escapes through the wire-gauze into the outer drum, while the crystals are rapidly dried.

† We can, and should, test all the candy we purchase by putting a small piece in a glass of water. Whatever settles to the bottom and remains undissolved is an adulteration.

Confectionery is often colored by dangerous poisons, so that prudence forbids the use of any colored candy. Licorice drops are frequently only the poorest brown sugar, terra alba, and a flavoring of licorice to make the unwholesome mixture palatable. Gum-drops are made, not from gum arabic, but generally of a species of glue manufactured out of hoofs, parings of hides, etc. However repugnant it may appear, this substance is perfectly clean and wholesome. Rock candy is formed by suspending threads in a strong solution of sugar. It crystallizes upon the rough surface in large, six-sided prisms.

Caramel, familiarly called burnt sugar, is formed whenever sugar is heated above its melting point, as when sweetmeats boil over on the stove; H_2O is lost, and C remains in excess. It is used by confectioners and for coloring liquors.

Grape - Sugar ($C_6H_{12}O_6$), *Dextrose*, is found in honey, figs, and many kinds of fruit. Its sweetening power is about three fifths that of cane-sugar.

Sugar from Starch.—The difference in the constitution of starch and grape-sugar is only H_2O. By boiling corn or potato-starch with dilute H_2SO_4, it is transformed into dextrose. The solution, evaporated to a syrup, is known commercially as "glucose," "mixing syrup," etc. When evaporated to dryness, the product is known as "grape-sugar."*

---

* Saw-dust, paper, and even rags, can in the same way be converted into sugar. Indeed, Professor Pepper speaks of seeing some made out of an old shirt. Wonderful beyond our comprehension is that chemical force which can transform a cast-off garment into a substance which will delight the palate. (The transformation of rags into paper is not a *chemical* one.)

"Candied Jellies, Preserves, Etc."—The sugar of many kinds of ripe fruits consists of grape or cane sugar, mixed with fruit-sugar. The latter changes gradually into grape-sugar and crystallizes, as in honey, dried figs, etc.*

Necessity of Organization.—We have found many elements which are necessary to the growth of our bodies, but still we can not live upon them. We need phosphorus, but we can not eat it, for it is a deadly poison. We need Fe, but it would make a most unsavory diet. We need CaO, but it would corrode our flesh. We need H, but it must be combined with O, as in H_2O, to be of any value to us. We need C, but charcoal would form a very indigestible food. If we were shut up in a room with all the elements of nature, we not only could not combine them so as to produce those organic substances necessary to our life and comfort, but we should actually die of starvation. We thus find that the mineral matter must be organized before we can use it to advantage.

Plants Organize Matter.—We have seen that in the plant the sunbeam decomposes CO_2, and returns to the air the life-giving O; that we can not create energy ourselves, or draw it direct from the sun,

Thus the chemist faintly imitates nature, which, ever out of waste and refuse, springs afresh. The fair petals of the lily rest upon the black mud of the swamp, and the products of decay come back to us in objects of use and forms of beauty.

* Fruit-sugar is isomeric with grape-sugar, but is much sweeter. The latter, as it is noted for its right-handed rotation of the plane of polarized light, is called *dextrose* (*dextra*, right), and the former, from its left-handed rotation, *lævulose* (*lævus*, left). (See "Physics," p. 170.)

but must take that which the plant has hoarded for us. We shall now find that, in addition, the plant changes *inorganic* matter to *organic*. It takes up the elements we need for our growth and for use in the arts, and combines them into plant-products, such as wood, starch, sugar, etc. We are thus dependent upon the vegetable world for the grand staples of commerce and of luxury—all that we eat, drink or wear. Each tiny leaf, every tree and shrub, every spire of grass even, is working constantly for us. The earth was once a burnt body—the cinders of the vast fire amid which it had its origin. (See "Geology," p. 17.) Every organized substance now on its surface has been rescued from the grasp of O by the plants.

THE AROMATIC COMPOUNDS.

THE name of this large and important group of compounds is due to the fact that many of its members occur in balsams, resins and essential oils which have an aromatic odor. The simplest aromatic hydrocarbon is *benzol*, or *benzene*, C_6H_6, and the other hydrocarbons and compounds of the group are derivatives of this, formed by the replacement of its H by elements or groups.

Benzene, C_6H_6, is a product of the distillation of coal-tar, obtained in gas-making (see p. 72). It is a colorless oil, which is a good solvent for guttapercha, caoutchouc, and fats. It is lighter than water, boils at 80.5° C., and is chiefly used for the preparation of its derivatives, many of which are of the greatest importance.

Nitro-Benzene, $C_6H_5NO_2$, is made by the action of nitric acid on benzene. It is a heavy, oily liquid, with an odor like that of bitter almonds. It is sometimes called essence of mirbane, and is used in scenting soap and in perfumery, but is chiefly valuable as the source of aniline, from which are prepared the celebrated coal-tar dyes.—*Example:* Mauve, magenta, etc. Who but a chemist would have searched in black, sticky coal-tar for these rainbow-tints, the stored-up sunshine of the carboniferous age!

Aniline, $C_6H_5NH_2$, is formed by the action of nascent H on nitro-benzene. On a large scale, it is made by mixing the nitro-benzene with iron filings and HCl. It is a colorless liquid, which becomes rapidly colored on exposure to the air. It is a strong base, uniting with nearly all acids to form crystalline salts. In this characteristic it resembles ammonia, NH_3, and it may be looked upon as ammonia in which an H atom has been replaced by the organic radical *phenyl*, C_6H_5. By treatment with oxidizing agents, aniline yields a great variety of derivatives, among them the exceedingly valuable aniline dyes.*

* "In 1856, Mr. Perkin, while experimenting with aniline in hopes of making quinine, treated it with potassium bichromate. He did not succeed in his attempt, but he obtained a beautiful purple dye, which was soon introduced to commerce under the name of *mauve.* A host of imitators at once sought to obtain the color without using potassium bichromate. As the only use of the latter was to oxidize the aniline, they reasoned that they might use any other oxidizing agent. Arsenic, among other substances, was tried, but, instead of a purple, the red known as *magenta* was the result.

Aniline was discovered in 1826, among the products of the dry distillation of indigo, and received its name from *anil*, the Portuguese term for indigo.

Phenol, C_6H_5OH, *Carbolic Acid*,* is noted for its

The coloring matter, however, does not contain any arsenic; being a salt of a base called *rosaniline*. Rosaniline itself is colorless, and reveals its magnificent tints only in its compounds. 'The crystals of its salts exhibit by reflected light the metallic green color of beetles' wings, but are of a deep red color when seen by transmitted light.' Magenta is manufactured on an enormous scale in England, more as a substance from which to obtain other dyes than for direct use in dyeing. A single firm produces twelve tons a week. The quantity of magenta furnished by one hundred pounds of coal is very small; but this is compensated for by its intense coloring power, since it will dye a quantity of wool nearly equal in weight to the coal. In making magenta on the large scale, there are considerable quantities of residual products. These, of course, have been examined with a view to further profit, and the result has been the discovery of a beautiful orange color called *phosphine*. This is much used to produce scarlet, by first dyeing the silk or wool with magenta, and then passing it through a bath of phosphine. By treating magenta with aniline, a beautiful blue is obtained. This is insoluble in water, but is rendered soluble exactly as indigo is, by treating it with sulphuric acid. Another curious dye formed from aniline is known as *Nicholson's blue*. This is completely discolorized by alkalies, and the color is restored by acids. In dyeing with it, the silk or wool is first immersed in a colorless solution of the dye, and then dipped into dilute sulphuric acid, when the blue is at once developed. If magenta is heated with iodide of ethyl or methyl, an excess of the iodide being employed, a most beautiful green is the result. If, however, this green is heated sufficiently to drive off the excess of iodide, a violet color is the result; so that it will not do for ladies wearing dresses dyed with this green to sit too near the fire. After all the coloring matter has been extracted from the aniline, a residue remains which has an intense black color, and is largely used for making printing-ink. Very few of the aniline colors, when in powder, give a person any idea of the color which they will produce when moistened. Magenta, for instance, when dry, is a beautiful green, with a bronze-like luster. It is a pretty experiment to coat a sheet of glass with one of these colors, which is readily done by dissolving in alcohol (Hofmann's violet being the best), and allowing a film of it to evaporate on the glass. When seen by transmitted light it is of a beautiful violet, but with reflected light it displays a tint rivaling in brilliancy the tail of a peacock."—*Boston Journal of Chemistry*.

* The acid may be considered as the hydroxide of the radical phenyl, and hence is sometimes called *phenyl alcohol*.

antiseptic and disinfecting properties. It is one of the products of coal-tar distillation, and forms white crystals. It is very poisonous. By heating it with HNO_3, $C_6H_2(NO_3)_3OH$, *picric acid* is formed. This colors a rich yellow, and is a very popular silk dye. It forms salts by the replacement of the H of the OH group, *e. g.* $C_6H_2(NO_2)_3.OK$. The picrates are yellow, explosive salts. Potassium picrate is used in making certain explosives.

Among the other products obtained by distilling coal-tar are: *Coal-tar Naphtha*; this is a volatile, limpid oil, with a peculiar odor, and generally a light straw color. It is composed of several hydrocarbons, and is very inflammable. *Naphthalene* is a crystalline solid occurring in beautiful pearly scales. It is especially abundant in dead-oil, and may be formed by passing olefiant gas or benzol through red-hot tubes. *Anthracene* accompanies naphthalene in the latter part of its distillation. It is also a white solid. It is of interest since the coloring principle of madder—alizarine—has been made from it. *Dead-oil* is used for preserving timber; as a cement for roofs and walls; for oiling machinery, etc.

The Acids of this group contain the carboxyl group —CO.OH, like the acids of the paraffine series. Among them are *benzoic acid*, $C_6H_5CO.OH$; *salicylic acid*, $C_6H_4(OH)CO.OH$, both of which are used in medicine, the latter and its salts being especially important.

Benzoic Aldehyde, C_6H_5CHO, is the fragrant oil of bitter almonds, and methyl salicylate, $C_6H_4(OH)$

$CO.OCH_3$ (an ethereal salt), is the natural oil of the wintergreen.

Toluene, $C_6H_5CH_3$ (methyl benzene), is, next to benzene, the most important of the hydrocarbons of the aromatic group. Like benzene, it yields, when treated with nitric acid, a nitro-compound, *nitro-toluene*, $C_6H_4NO_2CH_3$, which is reduced by nascent H to toluidine, $C_6H_4NH_2CH_3$. Toluene is always present in commercial benzene, and hence common aniline contains some toluidine, which is of importance in the making of aniline dyes.

THE TERPENES AND CAM-PHORS.

The Volatile or Essential Oils.—The volatile oils, unlike the fixed, make no soaps, and dissolve readily in alcohol or ether. Their solution in alcohol forms an essence.

Sources. — The volatile or essential oils are of vegetable origin. They are found in the petals of a flower, as the violet; in the seed, as caraway; in the leaves, as mint, or in the root, as sassafras. Some-times several kinds of oil are obtained from different parts of the same plant.—*Example:* In the orange tree, the flower, leaves, and rind of the fruit furnish each its own variety. The perfume of flowers is produced by these volatile oils; but how slight a quantity is present may be inferred from the statement that "one hundred pounds of fresh roses will give scarcely a quarter of an ounce of Attar of Roses."

Preparation.—In the peppermint and many others, the plant is distilled with water. The oils pass over with the steam, and are condensed in a cooler connected with the "Mint Still." The oil floats on the surface of the condensed water, and may be removed. A small portion, however, remains mingled with the latter, which thus acquires its peculiar taste and odor, constituting what is termed a "perfumed water." In some flowers, as the violet, jasmine, etc., the perfume is too delicate to be collected in this manner. They are, therefore, laid between woolen cloths saturated with some fixed oil. This absorbs the essential oil, which is then dissolved by alcohol. The oil of lemon or orange, is obtained from the rind of the fruit by expression, or by digesting in alcohol.

Composition.—$C_{10}H_{16}$ is the common formula of a large number of these oils. Thus the oils of lemon, cloves, juniper, birch, black pepper, ginger, bergamot, turpentine, cubebs, oranges, etc., nearly twenty in all, are isomeric.

Oil of Turpentine ($C_{10}H_{16}$), is a type of this group. It is made by distilling pitch. It is generally called *spirits of turpentine.* It is highly inflammable, and, owing to the excess of C, burns with a great smoke. Turpentine is used in making varnishes and in medicine. By the union of two atoms of its H with an atom of the O of the air, to form H_2O, it is converted into resin.* *Camphene* is obtained by repeated distillation of turpentine. *Burning-fluid* is a mixture of camphene and alcohol. In the heat of the burning H of the latter, the C of the former is consumed, and this produces a bright light. The tendency of camphene to smoke is thus diminished, and the illuminating power increased. By the action of HCl on turpentine or oil of lemons, an "artificial camphor" is produced, which much resembles common camphor.

Camphor ($C_{10}H_{16}O$) is obtained by distilling chips of the camphor-tree and its roots with water, and condensing the vapors on rice-straw. It is purified by sublimation. When kept in a bottle, it vaporizes, and its delicate crystals collect on the side toward the light. Taken internally, except in small doses, it is a virulent poison. Its solution in alcohol is called "spirits of camphor." If H_2O be added to this, the camphor will be precipitated as a flour-like powder.†

The Resins and Balsams.—The resins are generally formed from the essential oils by a slow oxidation.—*Example:* Turpentine, as we have just seen, is changed to rosin, a resinous substance. If the resin is dissolved in some essential oil, it is

---

* In this way, the turpentine around the nozzle of a bottle in which it is kept becomes first sticky, and then resinous. Old oil should not be taken to remove grease spots, as, while it will remove one, it will leave another of its own.

† Though camphor-gum is powdered with difficulty, a few drops of alcohol will remove all trouble. When small particles of powdered camphor are thrown on water free from grease, each fragment begins to dissolve with a remarkable gyratory motion, which is instantly checked by a drop of an essential oil allowed to fall upon the surface of the liquid.

called a balsam.—*Example:* Pitch is a balsam, since
by distillation it is separated into rosin and turpentine. They generally exude from incisions in trees
and shrubs, in the form of a balsam, which oxidizes
on exposure to the air, and becomes a resin.—*Example:* Spruce gum. The resins are translucent or
transparent, brittle, insoluble in H_2O, but soluble in
ether, alcohol, or any volatile oil, and form varnishes.
They are non-conductors of electricity, and burn with
much smoke. They do not decay, and indeed, have
the power of preserving other substances.*

Rosin constitutes about 75 per cent. of *pitch*, a
resinous substance which exudes from incisions made
in the trunks of certain species of pine. It is used
in making soaps, to increase friction in violin-bows
and the cords of clock-weights, and in soldering.

Lac exudes from the ficus-tree of the East Indies.
An insect punctures the bark, and the juice flows
out over the insect, which works it into cells in
which to deposit its eggs. The dried gum incrusting
the twigs is called *stick-lac;* when removed from the
wood, *seed-lac;* when melted and strained, *shellac.*
The liquefied resin is dropped upon large leaves, and
so cools in broad, thin pieces. *Sealing-wax* is made
of shellac and Venice turpentine; vermilion or red
lead being added to give the red color. Shellac is
much used in making varnishes.

Gum Benzoin also exudes from a tree in the East

* For this reason they were used in embalming the bodies of the ancient
Egyptians, which, after the lapse of two thousand years, are yet found dried
into mummies in their mammoth tombs—the Pyramids.

Indies. It is a source of benzoic acid. It is used in fumigation and in cosmetics, and on account of its fragrant odor is burnt as incense.*

Amber is a fossil resin which has exuded in some past age of the world's history from trees now extinct. It is sometimes found containing various insects perfectly preserved, which were without doubt entangled in the mass while it was yet soft. These are so beautifully embalmed in this transparent glass that they give us a good idea of the insect life of that age. Amber is cast up by the sea, principally along the shores of the Baltic; although it is also found in beds of lignite. It is commonly translucent, and susceptible of a high polish. It is used for ornaments, mouth-pieces, necklaces, buttons, etc.; and is an ingredient of some varnishes.

Caoutchouc, or *India-rubber*, exudes from certain trees in South America as a milky juice.† The solvents of rubber are ether, naphtha, turpentine, chloroform, carbon disulphide, etc. It melts, but does not become solid on cooling. Freshly-cut sur-

---

* Place some green sprigs under a glass receiver, and at the bottom a hot iron, on which sprinkle a little benzoic acid. It will sublime and collect in beautifully delicate crystals on the green leaves above, making a perfect illustration of winter frost-work.

† The globules of rubber are suspended in it as butter is in milk. The tree, it is said, yields about a gill per day from each incision made. A little clay cup is placed underneath, from which the juice is collected and poured over clay or wooden patterns in successive layers as it dries. To hasten the process it is carried on over large open fires, the smoke of which gives to the rubber its black color; when pure, it is almost white. When nearly hard, the rubber will receive any fanciful design which may be marked upon it with a pointed stick. The natives often form the clay into odd shapes, as bottles, images, etc., and the rubber is sometimes exported in these uncouth forms.

faces readily cohere ; this property, together with its power of resisting most re-agents, renders it invaluable to the chemist in making flexible joints and tubes. · "It loses its elastic power when stretched for a long time, but recovers it on being heated. In the manufacture of rubber goods for suspenders, etc., the rubber thread is drawn over bobbins, and left for some days until it becomes inelastic. In this state it is woven, after which a hot wheel is rolled over the cloth to restore the elasticity."

Vulcanized Rubber is made by heating caoutchouc with a small amount of sulphur. This constituted Goodyear's original patent.* It is less liable to be hardened by cold, or softened by heat, and admits of many uses to which common rubber would be entirely unsuited. If sulphurized rubber be heated to a high temperature, it becomes a hard, brittle, black solid (vulcanite or ebonite), capable of

---

* Mr. Goodyear had been experimenting to find some way of rendering rubber insensible to heat and cold. It is said that one day, while talking with a friend, he happened to drop a bit of S in a pot of melted rubber. By one of those happy intuitions which seem to come only to men of genius, he watched the process, and to his amazement found that, while the appearance of the rubber was the same—elastic, odorous, and tasteless— its stickiness was gone, and it had gained the properties he so much desired. He immediately took out a patent in this country, and sailed for England, where, instead of securing his secret by a similar patent, he offered to sell it for £10,000. Charles Hancock, with whom he had been corresponding for several years, and who had been engaged in similar experimenting, resolved to discover it himself. He shut himself up in his laboratory, and went to work. Disheartening failures marked every attempt. At last he tried S. At first, he did not succeed : but, persevering, he finally saw, amid the stifling fumes of brimstone, the soft rubber metamorphosed into the vulcanized caoutchouc. He, too, was possessed of the secret, and, taking out a patent, reaped the reward of his patient labor.

a high polish, which is used for knife-handles, combs, buttons, etc.

Gutta-percha resembles caoutchouc in its source, preparation, and appearance. It softens in hot water, and can then be molded like wax. When cooled, it assumes its original solidity. It is extensively used in taking impressions of medals, etc.

THE ALKALOIDS.

THE alkaloids, or organic bases, as they are called, are the bases of true salts found in many plants. They dissolve very slightly in H_2O, but freely in alcohol. They have a bitter taste, and rank among the most fearful poisons and valuable medicines. All the alkaloids contain N.*

Opium is the dried juice of the poppy plant, which is extensively cultivated in Turkey for the sake of this product. Workmen pass along the rows soon after the flowers have fallen off, cutting slightly each capsule. From these incisions a milky juice exudes, and collects in little tears. These are gathered, and wrapped in leaves for the market. Opium contains some seventeen different alkaloids in combination with at least two acids. In small doses, opium is a

---

* A convenient antidote is tannin, which forms with nearly all of them insoluble curdy tannates. Almost any liquid containing it is of value—as strong, green tea. This is also of use, as it tends to keep the patient awake, the great necessity in the case of a narcotic poison.

sedative medicine; in larger ones, a narcotic poison. *Laudanum* is the tincture of opium; and *paregoric*, a camphorated tincture flavored with aromatics.

Opium-eating.—Opium produces a powerful influence on the nervous system. It stimulates the brain and excites the imagination to a wonderful pitch of intensity. The dreams of the opium-eater are said to be vivid and fantastic beyond description. The dose must, however, be gradually increased to repeat the effect, and the result is most disastrous. The nervous system becomes deranged, and no relief can be secured save by a fresh resort to this baneful drug.* Labor becomes irksome, ordinary food distasteful, and racking pains torment the body.

Morphine (*Morpheus*, the god of sleep) is the chief narcotic principle of opium, and like it is used to alleviate pain and produce sleep. It is usually given as a sulphate or chloride.

Quinine is prepared from Peruvian bark. A tinct-

---

* In time, the whole system becomes so impregnated with it that even large additional doses fail to produce the delightful effect which at first so fascinated the victim. Then, while acting upon the nerves, it set free a vast amount of vitality and energy, but now it has satisfied itself. The subtle alkaloid has affected the tissues and coatings of the entire internal organism. If, resolutely, one summons his enfeebled will, and commences the conflict, an agony of endurance, which defies all description, is before him. The whole body must be reorganized. If, too weak to attempt so terrible a struggle, he continues the use of the fatal drug, he moves on directly to his fate—the opium-eater's grave. Paregoric, laudanum, morphine, and the different preparations of opium, are, in almost every case, taken first as a sedative from pain or fatiguing labor, with no thought of becoming addicted to their use. But so insinuating is it that the victim forms the habit ere he is aware, and knows he is a slave only when he attempts to cease the customary dose. No person can be too careful in beginning the use of a narcotic whose influence is liable to become so destructive.

ure of the bark, or sulphate of quinia, is employed in medicine in cases of fever and ague, and other periodic diseases, and also as a tonic.

Nicotine is the active principle of the tobacco plant, of which it forms from 2 to 8 per cent. It is volatile, and passes off in the smoke. A drop will kill a large dog. It probably produces many of the ill effects which follow the use of tobacco.

Strychnine is prepared from the nux vomica and the St. Ignatius bean. It is also a constituent of the celebrated upas poison.* "It is so intensely bitter that one grain will impart a flavor to twenty-five gallons of water. One thirtieth of a grain has killed a dog in thirty seconds, while half a grain is fatal to man."

The *Chromatic Test*, as it is called, consists in placing on a clean porcelain plate a drop of the suspected liquid, a drop of H_2SO_4, and a crystal of potassium bichromate. Mix the three very slowly with a clean glass rod. If there be any strychnine present, it will change the color into a beautiful violet tint, passing into a pale rose.†

---

* "The 'woorara,' with which the South American Indians poison their arrows, is a variety of strychnine. This is so deadly that the scratch of a needle dipped in it will produce death; yet it may be swallowed with impunity."—MILLER.

† Strychnine is the only poison, except brucine (and that also is extracted from nux vomica), that produces tetanus, or lock-jaw. This symptom proves to the physician that death has been caused by this alkaloid. To exhibit the effect of the poison a frog may be brought into the court-room, and made to show its action. So sensitive is this little animal that a few drops of oil, containing only a hundred-thousandth of a grain of strychnine, will instantly throw it into a rigid lock-jaw, in which it is incapable of the least motion.

Caffeine and Theine constitute the active principle of tea* and coffee,† and are isomeric. They crystallize in long, flexible, silky needles. In addition, tea contains from 13 to 18 per cent. of a form of tannin (see p. 237, note), about 22 per cent. of extractive matter, some coloring substances, and a volatile oil which gives to it its aromatic odor and taste. Coffee contains about 14 per cent. of oil and fat, and also an essential oil which is developed in roasting, and is very volatile, so that it will soon escape unless the coffee be kept tightly covered.

* *Tea-raising.*—Tea-plants resemble in some respects the low whortleberry bush. They are raised in rows, three to five in a hill, very much as corn is with us, but they are not allowed to grow over one and a half feet high. The medium-sized leaves are picked by hand, the largest ones being left to favor the growth of the bushes. Each little hill or clump will furnish from three to five ounces of green leaves, or about one ounce of tea, in the course of the season. The leaves are first wilted in the sun, then trodden in baskets by barefooted men to break the stems, next rolled by the hands into a spiral shape, then left in a heap to heat again, and finally dried for the market. This constitutes *Black Tea*, and the color would be produced in any leaves left thus to wilt and heat in heaps in the open air. The Chinese always drink this kind of tea. They use no milk or sugar, and prepare it, not by steeping, but by pouring hot water on the tea and allowing it to stand for a few minutes. Whenever a friend calls on a Chinaman, common politeness requires that a cup of tea be immediately offered him.

Green Tea is prepared like black, except that it is not allowed to wilt or heat, and is quickly dried over a fire. It is also very frequently, if not always, colored—cheap black teas and leaves of other plants being added in large quantities. In this country, damaged teas and the "grounds" left at hotels are re-rolled, highly colored, packed in old tea-chests, and sent out as new teas. Certain varieties of black tea even receive a coating of black-lead to make them shiny.

† Coffee is the seed of the coffee plant, a native of the tropics. The plant is very prolific, remaining in flower eight months of the year and usually producing three harvests annually. The fruit resembles the cherry, but contains two seeds or "beans" instead of a single stone, inclosed in a thick leathery skin. The drying of the coffee is a most important process. A shower of rain will discolor the bean and depreciate its value.

DYES AND DYEING.

MANY of the organic coloring principles are of vegetable origin. They are found in the roots, wood, bark, flowers, and seeds of plants.

Dyeing.—Very few of the colors have such an affinity for the fibers of the cloth that they will not wash out. Those which, like indigo, will dye directly, are called *substantive* colors. But the majority are *adjective* colors, which require a third substance having an attraction for both the coloring matter and the cloth, to hold them together. Such substances are called *mordants* (*mordeo*, to bite), because they bind the dye in the cloth, thus making a "fast color." The most common mordants are alum, salts of tin and of iron. In dyeing, the cloth is first dipped into a solution of the mordant, and then into one of the dye-stuff. The mordant, by means of a stamp, may be applied to the cloth in the form of a pattern, and, when it is afterward washed, the color will be removed, except where the mordant fixed it in the printed figure. The same dye will produce different colors by a change of mordants.—*Example:* Madder, with iron, gives a fine purple ; with alum, a pink ; and with iron and alum, a chocolate. This principle lies at the basis of dyeing "prints."*

---

* A calico printing-machine is very complex. The cloth passes between a series of rollers, upon which the corresponding mordant is put, as ink is on type. A single machine sometimes prints from twenty sets of rollers ; yet each impression follows the other so accurately that, when the cloth has passed through, the entire pattern is printed upon it, with the different mordants, more perfectly than any painter could do it, and so rapidly that

Coloring Substances.—*Madder* is the root of a plant found in the East, and extensively cultivated elsewhere. When first dug, it is yellow, but by exposure it becomes red. It is used in dyeing the brilliant Turkey-red. The coloring principle, which is named *alizarin*, is identical with that derived from anthracene, a hydrocarbon found in coal-tar (see p. 224), and is now made in large quantities from that source. *Cochineal* is a dried insect that in life feeds upon a species of cactus in Central America. The coloring matter is called *Carmine*. It yields the brightest crimson and purple dyes.* *Brazil-wood* furnishes a red which is not very permanent. It is used for making red ink. The *indigo* of commerce is obtained from a bushy plant found in the East Indies. By fermenting for some days in vats of water, the coloring matter is developed. Reducing agents change indigo into a soluble and colorless substance by the absorption of H.† It is then called "white indigo." In this form it is extensively used in dyeing. The cloth becomes permanently colored on

a mile of cloth has been printed, with four mordants, in an hour. The cloth, when it leaves the printing-machine, though stamped with the mordants in the form of the figure, betrays nothing of the real design until after being dipped in the dye, which, acting on the different mordants, brings out the desired colors. The print is now washed, glazed, and fitted for the market.

* The purple of which we read in ancient writings was a secret with the Tyrians. King Huram, we learn, sent a workman to Solomon skilled in this art. The dye was obtained from a shell-fish that was found on the coast of Phœnicia. Each animal yielded a tiny drop of the precious liquid. "A yard of cloth dipped twice in this costly dye was worth $150."

† Place a little powdered indigo in a test-tube of H_2O, and add zinc filings and caustic soda. On heating, the indigo becomes colorless. If it is now exposed to the air in a saucer, it will turn blue again.

exposure to the air, when the insoluble blue indigo is formed in its fibers.* *Logwood* is so named because imported in logs. It is the heart of a South and Central American tree. With a mordant of iron, it dyes black. *Litmus* is obtained from a variety of lichens common along the southern coast of Europe. Its juice is colorless, but by the action of water, air, and NH_3, it assumes a rich purple blue. *Leaf-green* (chlorophyl), as found in plants, is a resinous substance containing several coloring matters. It seems to be developed by the action of the sunbeam. Plants removed from a dark cellar to the sunlight, rapidly turn green.

Tannic Acid, *tannin* ($C_{14}H_{10}O_9$), is found in the leaf and bark of trees.†—*Example:* Oak, hemlock. Nut-galls are excrescences which are formed by the puncture of an insect on the leaves of a certain species of oak. Tannin has an astringent taste, is soluble in water, and hardens albuminous substances, as gelatin.

Tanning.—After the hair has been removed from the skins by milk of lime, they are soaked for days, the best kinds for months, in vats full of water and ground oak or hemlock bark (tan-bark). The tannic acid of the bark is dissolved, and, entering the pores of the skin, unites with the gelatin, forming a hard,

* One of the recent triumphs of synthetical chemistry has been the artificial preparation of indigo.

† This astringent principle is widely diffused. There are several compounds which possess similar properties, yet differ in chemical composition. The tannin of the oak is called *quercitannic acid;* that of nut-galls, *gallotannic acid;* that of tea, *theitannic,* and that of coffee, *caffeotannic acid.*

insoluble compound which is the basis of leather. Leather is blackened by washing the hide on one side with a solution of copperas. The tannic acid unites with the iron, forming a tannate of iron—an ink. In the same way, drops of tea on a knife-blade stain it black.

Ink is made by adding a solution of nut-galls to one of copperas. The iron tannate thus formed has a pale blue-black color, as in the best writing-fluids; by exposure to the air, the Fe absorbs more O, the ink darkening in color until it is a deep black. Gum is added to thicken and regulate the flow of the fluid from the pen. Creosote, or corrosive sublimate, is used to prevent moldiness. Steel pens are corroded by the free H_2SO_4 contained in the ink, but gold pens are not affected by it.[*]

Gallic Acid ($C_7H_6O_5$) is best prepared from nut-galls, by fermentation of the tannic acid which they contain. *Pyrogallic acid* can be obtained by the sublimation of gallic or gallotannic acid. It is extensively used in photography for the purpose of developing the latent image in the collodion film after exposure to the action of the light. (See p. 172.)

Linseed Oil is a *drying oil*, as it is termed—*i. e.*, it absorbs O from the air, and hardens by exposure.

---

[*] The following is an instructive experiment, illustrating the manner of making ink, of removing stains with oxalic acid, and also the relative strength of the acids and alkalies: Take a large test-tube, and add the following re-agents in solution, *cautiously, drop by drop*, watching the result and explaining the reactions: 1, iron sulphate (*copperas*); 2, tannic acid (*tannin*); 3, oxalic acid; 4, sodium carbonate (*sal-soda*); 5, hydrochloric acid (*muriatic*); 6, ammonia (*hartshorn*); 7, nitric acid (*aquafortis*); 8, caustic potash; 9, sulphuric acid (*oil of vitriol.*)

It is expressed from flaxseed, which furnishes about one fifth of its weight in oil. *Boiled oil* is made by heating the crude oil with litharge, which entirely dissolves and greatly increases the drying property of the oil. Linseed oil is used in mixing paints and varnishes. *Putty* consists of linseed oil and whiting, well mixed. The chief ingredients of *printers' ink* are linseed oil, heated until it becomes thick and viscid, and lamp-black.

THE ALBUMINOUS BODIES.

These are albumin, casein, and fibrin. Owing to the complexity of their composition, no satisfactory formula can be assigned to them. The molecule of albumin has been stated as $C_{72}H_{110}N_{18}SO_{22}$, but it is very uncertain.*

Albumin is found nearly pure in the whites of eggs,† hence the name (*albus*, white). It exists as a liquid in the sap of plants, the humors of the eye, serum of the blood, etc.; and as a solid in the seeds of plants, and the nerves and brains of animals.‡

---

* Many chemists regard albumin, casein, fibrin, etc., as isomeric, and capable of being converted by the vital force one into the other. These bodies are sometimes called *Protein* (*protos*, first) on the supposition that they were derived from a single azotized principle named protein.

† Strange to say, "the venom of the rattlesnake is isomeric with the 'the whites of eggs.'"

‡ This principle is of very great importance, as albumin may thus be carried by the blood through the system, but when once deposited it can not be dissolved and washed away again.

Properties.—It is soluble in cold, but insoluble in hot H_2O. At a temperature of about 140° F., it coagulates. This change we always see in the cooking of eggs; yet nothing is known of its cause. Alcohol, corrosive sublimate, acids, creosote, and solutions of copper, lead, silver, etc., have the power to coagulate albumin. In cases of poisoning by these substances, the white of eggs is therefore a valuable antidote, as it wraps them in an insoluble covering, and so protects the stomach.

Casein (*caseus*, cheese) is found in the curd of milk. In the presence of an acid it coagulates, and thus milk curdles after it sours. Rennet (the dried stomach of a calf) is used to coagulate milk in the process of cheese-making, but the cause of its action is not understood.

Milk is a natural emulsion, composed of exceedingly minute globules diffused through a transparent liquid. The globules consist of a thin envelope of casein filled with butter. Being a trifle lighter than H_2O, they rise to the surface as cream. Churning breaks these coverings, and gathers the butter into a mass. Milk contains some sugar, which, by a peculiar change termed "lactic fermentation," is converted into lactic acid. The casein seems to act as a ferment in hastening this oxidation, and, by its decay, produces the offensive odor. In the "souring" of milk, the milk-sugar ($C_{12}H_{24}O_{11}$) disappears, and

FIG. 73.

Milk under the Microscope.

lactic acid ($C_3H_6O_3$) gradually takes its place. It is an excellent illustration of a complex molecule breaking up into simple ones.

Fibrin constitutes chiefly the fibrous portion of the muscles. If a piece of lean beef be washed in clean H_2O until all the red color disappears, the mass of white tissue which will remain is called *fibrin*. Like albumin, it exists in two forms—as a liquid in the blood, and as a solid in the flesh. The clotting of blood is due to the coagulation of the fibrin. (See "Physiology," p. 108.)

Fig. 74.

Fibrin, or Muscle.

Vegetable Albuminoids.—Vegetables contain substances which are scarcely to be distinguished from the albuminous bodies derived from animal sources. If wheat flour be made into a dough, and then kneaded in water until the soluble portion is washed away, the tough, sticky mass which will remain is called *gluten*. It is a nitrogenous substance, allied to albumin. It exists most abundantly in the bran of cereal grains.

By treating peas as we do potatoes in forming starch, and then adding a little acid to the water which is left after the starch settles, an albuminous substance is deposited, which is thought to be identical with casein. The Chinese use it largely for cheese. It is found abundantly in the seeds of peas, beans, etc., and is termed *legumin*.

Putrefaction.—Owing to the complex structure of albuminous substances, and the presence of N, they readily oxidize, and form new and simple compounds. This breaking up of the organic structure is called putrefaction, and is but a special kind of fermentation. The activity of the ferment probably explains the danger physicians incur in dissecting dead bodies. The least portion of the decomposing matter entering the flesh, through a scratch, is liable to be fatal. The absence of H_2O retards chemical change, and, therefore, meats, apples, etc., are preserved by drying.* Salt acts by hardening the albumin, by absorbing the juice of the meat, and by covering as brine, and so warding off the attacking O; but as it dissolves some of the salts and other valuable elements, it makes the meat less nutritious.

Gelatin.—Hot water dissolves a substance from animal membranes, skin, tendons, and bones,† which,

* The cold also protects from chemical change. The bodies of mammoths have been found in the frozen soil of the Arctic regions so perfectly preserved that the dogs ate the flesh. How long the animals had been there we can not tell, but certainly for ages. In 1861 the mangled remains of three guides were found at the foot of the Glacier de Boisson, in Switzerland. They had been lost in an avalanche on the plateau of Mont Blanc, forty-one years before.

† ANALYSIS. (*Berzelius.*)

Gelatin..................................	32.17
Blood-vessels	1.13
Phosphate of lime.....................	51.04
Carbonate of lime....................	11.30
Fluoride of calcium...................	2.00
Phosphate of magnesia..............	1.16
Chloride of sodium...................	1.20
	100.00

Bones consist of organic and mineral matter combined. By soaking a

on cooling, forms a yielding, tremulous mass called gelatin. In calves-foot jelly, soups, etc., it is well known.* *Glue* is a gelatin made from bones, hoofs, horns, etc., by boiling in H_2O, and then evaporating the solution. *Isinglass* is a very pure gelatin, obtained from the air-bladders of the cod, sturgeon, and other fish. *Size* is a gelatin prepared from the parings of parchment. It is used for sizing paper in order to fill up the pores and prevent the ink from spreading, as it does on unsized or blotting-paper.

--- • ---

DOMESTIC CHEMISTRY.

In the chemistry of housekeeping there are some points not yet mentioned which may now be profitably discussed.

Making Bread.—Flour consists of starch, gluten, and a little dextrin and sugar.

The oily matter and the salts—of which there are from 1 to 2 per cent. in wheat—are contained mainly in the bran. The process of making the "sponge" is purely mechanical. When the sponge is set in a warm place to rise (as heat favors chemical

bone in HCl the mineral matter will all be dissolved, and the organic matter left in the original shape of the bone, but soft and pliable. If, instead, the bone be burned in the fire, the organic matter will be removed, and the mineral left white and porous. (See "Physiology," p. 20.)

* As an article of food it is of very little nutritive value. It may answer to dilute a stronger diet, but of itself does little to build up the body of an invalid. Beef-tea, even, is now thought to have little nourishing property, its principal office being to act as a stimulant.

change), the yeast, yeast-cake, or emptyings,* as the case may be, induce a rapid fermentation, converting the sugar into alcohol and CO_2. This gas is diffused through the mass, and, being retained by the tenacious dough, causes it to "rise"—*i. e.*, to swell and become porous. The next step includes the addition of flour, and a laborious process of "kneading." The latter diffuses the half-fermented sponge through the dough; it also breaks up into smaller ones the bubbles of gas entangled in the gluten, and makes the bread fine-grained. After a second rising, the dough is molded into loaves, which are set aside to perfect the fermentation. When they are finally placed in the oven, the heat expands the CO_2, and increases the porosity of the bread; the starch granules are broken up; and the alcohol is vaporized, and it and the H_2O partly driven off. The surface becomes dry and hard, and, losing a part of its chemically combined water, is partially converted into a substance allied to caramel, thus forming the crust.* If the temperature of the oven is right, the cells of the bread will have sufficient strength to

* Milk-emptyings are sometimes used in making bread. In this case the mixture of flour and milk, kept at a temperature of about "blood heat," rapidly develops yeast, which produces fermentation. If the heat is much above this, the plant will be killed, and the milk be merely turned to lactic acid. Oftentimes, too, the side of the dish, near the fire, may be warm enough to produce yeast, and to generate CO_2 and alcohol, while on the opposite side lactic acid is being formed. A uniform temperature is necessary, and this can best be obtained by placing the dish of emptyings in a kettle of warm water on the stove hearth.

† A shiny coat is given to the loaf ("rusk") by moistening the crust after the bread is baked, thus dissolving some of the dextrin, which is also contained in the crust. This quickly dries on returning it to the oven.

retain their form after the gas and vapors have escaped. If the heat is not sufficient, or if there is too much water in the dough, the CO_2 escapes, the cells, not being sufficiently hardened, collapse, and the bread is "slack-baked." If the oven is too hot, the crust forms too quickly over the surface of the loaf, preventing the escape of the CO_2, which accumulates at the center, making the loaf hollow.

Stale Bread.—New bread consists of about 45 per cent. water. In stale bread this disappears, but may be brought to view again by heating the loaf in a close tin vessel.

Aerated Bread is made in the following manner: Flour and salt are put in a revolving copper globe, into which H_2O, charged with CO_2, is admitted. When well mixed, a stop-cock is turned, and the dough is driven out by the elastic force of the gas, into pans ready for baking.

Sour Bread results from a neglect to arrest the first stage of the fermentation, thus allowing the second stage to commence, and acetic acid to be formed. The acid is neutralized by an alkali, as saleratus, or soda.

Griddle-cakes are raised by the addition of some ferment, as yeast; but the second, or acetic stage, is always reached. The "batter" then tastes sour, and is sweetened by saleratus or soda. The acetic acid combines with the metallic base, forming a harmless salt, which remains, while the CO_2 bubbles up through the batter, making it "light."

Raising Biscuit.—In raising biscuit or cake, soda and cream of tartar* are most commonly used. The CO_2 is set free, and, escaping as a gas, makes the dough porous, while the sodium and potassium tartrate (Rochelle salt) which is formed, remains. Ordinary "baking-powders" are merely cream of tartar and soda. A variety invented by Professor Horsford contains acid calcium phosphate (see note, p. 142); this, reacting upon the "soda," forms calcium and sodium phosphates, both of which are materials for bone-making.† Soda and HCl are also used in baking. By heat, these ingredients are resolved into H_2O, CO_2, and NaCl. The H_2O and CO_2 raise the bread, while the common salt seasons it. There is a difficulty in procuring pure acid, and in mixing the ingredients in their combining proportions. · Sal-volatile (ammonium sesquicarbonate, p. 137) is often used by bakers for raising cake. This should volatilize into two gases, NH_3 and CO_2, on the application of heat, but in practice a portion is commonly left hidden in the cake, and may be detected by the odor. Alum is often employed by bakers to whiten bread, and to improve the quality of bread made from inferior flour.

* Cream of tartar is often adulterated with plaster, lime, chalk, or flour. By dissolving in water, these impurities can be detected, as they form an insoluble precipitate; but in milk, as commonly used in cooking, they are not noticed.

† It is doubtful whether ordinary yeast-powders, or cream of tartar and soda, make as healthful food as the regular process of fermentation. There is frequently a portion of the powders left uncombined, and always a salt formed which may perhaps interfere with the action of the gastric juice. Sometimes, indeed, we find biscuit and cake yellow, and even spotted with bits of saleratus; yet, through a false economy, such food is too often "eaten to save it."

Toasting Bread.—By toasting, bread becomes much more digestible, as the starch is converted largely into dextrin, which is soluble. The charcoal which may be formed when the heat has disorganized the bread and driven off the water, also acts favorably on the stomach by absorbing in its pores noxious gases, as in "crust-coffee."

Cooking Potatoes.—A raw potato is indigestible, but by cooking the starch granules absorb the water of the potato, burst, and make it "mealy." If the potato contains more H_2O than the starch can imbibe, it is called "watery."

PRACTICAL QUESTIONS.

1. How would you prove the presence of tannin in tea?
2. How would you test for Fe in a solution?
3. Why can we settle coffee with an egg?
4. How would you show the presence of starch in a potato?
5. Why is starch stored in the seed of a plant?
6. Why are unbleached cotton goods dark colored?
7. Why do beans, rice, etc., swell when cooked?
8. Why does decaying wood darken?
9. How would you show that C exists in sugar?
10. Why do fruits lose their sweetness when over-ripe?
11. Why does maple-sap lose its sweetness when the leaf starts?
12. Should yeast cakes be allowed to freeze?
13. Why will wine sour if the bottle be not well corked?
14. Why can vinegar be made from sweetened water and brown paper?
15. Why should the vinegar-barrel be kept in a warm place?
16. Why does "scalding" check the "working" of preserves?
17. Is the oxalic acid in the pie-plant poisonous?
18. How may ink-stains be removed?
19. Why is leather black on only one side?
20. Why do drops of tea stain a knife-blade?
21. Why will not coffee stain it in the same way?
22. Why does writing-fluid darken on exposure to the air?

23. Why does ink corrode steel pens?
24. How does a bird obtain the $CaCO_3$ for its egg-shells?
25. Why will tallow make a harder soap than lard?
26. Why does new soap act on the hands more than old?
27. What is the shiny coat on certain leaves and fruits?
28. Why does turpentine burn with so much smoke?
29. Why is the nozzle of a turpentine bottle so sticky?
30. Why does kerosene give more light than alcohol?
31. What is the antidote to oxalic acid? Why?
32. Would you weaken camphor spirits with water?
33. What is the difference between rosin and resin?
34. Why does skim-milk look blue and new milk white?
35. Why does an ink-spot turn yellow after washing with soap?

CONCLUSION.

Chemistry of the Sunbeam.—The various plant-products of which we have spoken in Organic Chemistry, when burned, either in the body as food or in the air as fuel, give off heat. This was garnered in the plant while growing, and came from that great source of heat—the sun. Thus all vegetation contains the latent heat of the sunbeam, ready to be set free upon its own oxidation. The coal, even, derived as it is from ancient vegetation, hidden away in the earth, is thus a mine of reserved energy. Those black diamonds we use as fuel become, in the eye of science, crystallized sunbeams, fagots of energy, ready to impart to us at any moment the heat of some old Carboniferous day. A field of growing wheat reaches out its tiny arms, and, tangling in stalk and grain the heat of sultry mid-summer, retains it against the bleak December. The oil-well spouts not alone unsavory kerosene, but liquid sunbeams, the gathered store of a geologic age. As we warm ourselves by our fires, or sit and read by our oil and gas lights, how strange the thought that their light and heat streamed down upon the earth ages ago, were absorbed by the grotesque leaves of the old coal forests, and kept safely stored away by a Divine care in order to provide for our comfort! The present warmth of our bodies all came from the same source—the sun. It mostly fell in the sunbeams of last summer upon our gardens and fields, was preserved in the potatoes, cabbage, corn, etc., we have eaten as food, and to-day reappears as heat and motion. Every blow, every breath, and every step, are but transformations of solar rays, and can be estimated in sunshine.

The Sun the Source of Power.—The sun warms, enlivens, and animates the earth. In the laboratory of the leaf, he produces the most wonderful chemical changes. We see his handiwork in the building of the forest, the carpeting of the meadow, and the tinting of the rose. On the ladder of the sunbeam, water climbs to the sky, and falls again as rain. The very thunder of Niagara is but the sudden unbending of the spring that was first coiled by the sun in the evaporation from the ocean. Up to the sun, then, we trace all the hidden manifestations of power. Yet the energy that produces such intricate and wide-extended changes is only one twenty-three hundred millionth part of the tide that flows in every direction from this great central orb. But what is our sun itself save a twinkling star beside great suns like Sirius, Regulus, and Procyon, whose brilliancy in the far-off regions of space drowns our little sun as the dazzling light of day does the smoldering blaze of some wandering hunter?

Changes of Matter.—Chemical changes are taking place wherever we look—on land or sea. The hard granite crumbles and molders into dust. The stout oak draws in the air, and solidifies it; takes up the earth, and vitalizes it; changes all into its own structure, and proudly stands monarch of the forest. But in time its leaves turn yellow and sere; its branches crumble; itself totters, falls, and disappears. Our bodies seem to us comparatively stable, but, with the rock and the oak, they too pass away. All Nature is a torrent of ceaseless change. We are but parts of a grand system, and the elements we use are not our own. The water we drink and the food we eat to-day may have been used a thousand times before, and that by the vilest beggar or the lowest earth-worm. In Nature all is common, and no use is base. Those particles of matter we so fondly call our own, and decorate so carefully, a few months since may have dragged boats on the canal, or waved in the meadow as grass or corn.* From us they will pass on their ceaseless

---

* The truth that matter passes from the animal back to the vegetable, and from the vegetable to the animal kingdom again, received, not long since, a curious illustration. For the purpose of erecting a suitable monument in memory of Roger Williams, the founder of Rhode Island, his private burying-ground was searched for the graves of himself and wife. It was found that every thing had passed into oblivion. The shape of the coffins could only be traced by a black line of carbonaceous matter. The rusted hinges and nails, and a round wooden knot, alone remained in one grave; while a single lock of braided hair was found in the other. Near the graves stood an apple-tree. This had sent down two main roots into the very presence of the coffined dead. The larger root, pushing its way to the precise spot occupied by the skull of Roger Williams, had made a turn as if passing around it, and followed the direction of the backbone to the hips. Here it divided into two branches, sending one along each leg to the heel, when both turned upward to the toes. One of these roots formed a slight crook at the knee, which made the whole bear a striking resemblance to the human form. (These roots are now deposited in the museum

round to develop other forms of vegetation and life, whereby the same atom may freeze on arctic snows, bleach on torrid plains, be beauty in the poet's brain, strength in the blacksmith's arm, or beef on the butcher's block. Hamlet must have been somewhat more of a chemist than a madman when he gravely assured the king that "man may fish with the worm that hath eat of a king, and eat of the fish that hath fed of the worm."

Shakespeare expresses the same chemical thought when he again makes Hamlet say:

> " Imperious Cæsar, dead and turned to clay,
> Might stop a hole to keep the wind away.
> Oh! that the earth which kept the world in awe
> Should patch a wall to expel the winter's flaw!"

Or when he makes Ariel sing:

> " Full fathom five thy father lies:
> Of his bones are coral made;
> Those are pearls that were his eyes;
> Nothing of him that doth fade
> But doth suffer a sea change
> Into some thing rich and strange."

Life and Death are thus throughout nature commensurate with and companions of each other. Oxy-

---

of Brown University.) There were the graves, but their occupants had disappeared; the bones even had vanished. There stood the thief—the guilty apple-tree—caught in the very act of robbery. The spoliation was complete. The organic matter—the flesh, the bones, of Roger Williams—had passed into an apple-tree. The elements had been absorbed by the roots, transmuted into woody fiber, which could now be burned as fuel, or carved into ornaments; had bloomed into fragrant blossoms, which had delighted the eye of passers-by, and scattered the sweetest perfume of spring; more than that—had been converted into luscious fruit, which, from year to year, had been gathered and eaten. How pertinent, then, is the question, "Who ate Roger Williams?"

gen is the destroyer, and the sunbeam the builder. Oxygen tears down every living structure, and would bring all things to rest in ashes. The sunbeam re-invigorates, rebuilds, and rescues from the grasp of decay. Though they seem to be antagonists, oxygen and the sunbeam really work in harmony, and each supplements the labor of the other. Death alone makes life possible.

———

Thus we have traced some of the wonderful processes by which this world has been arranged to supply the varied wants of man. Wherever we have turned, we have found proofs of a Divine care planning, conforming, and directing to one universal end, while from the commonest things, and by the simplest means, the grandest results have been attained. Thus does Nature attest the sublime truth of Revelation, that in all, and through all, and over all, the Lord God omnipotent reigneth.

IV.
APPENDIX.

TABLE OF THE ELEMENTS.

Name.	Symbol.	Atomic Weight.	Name.	Symbol.	Atomic Weight
Aluminium	Al	27	Nickel	Ni	58
Antimony (Stibium)	Sb	120	Niobium	Nb	94
Arsenic	As	75	*Nitrogen*	N	14
Barium	Ba	137	Osmium	Os	199
Beryllium	Be	9	*Oxygen*	O	16
Bismuth	Bi	208	Palladium	Pd	106
Boron	B	11	*Phosphorus*	P	31
Bromine	Br	80	Platinum	Pt	194.2
Cadmium	Cd	112	Potassium (Kalium)	K	39
Cæsium	Cs	133	Rhodium	Rh	104
Calcium	Ca	40	Rubidium	Rb	85
Carbon	C	12	Ruthenium	Ru	104
Cerium	Ce	141	Samarium	Sm	150
Chlorine	Cl	35.5	Scandium	Sc	44
Chromium	Cr	52	*Selenium*	Se	79
Cobalt	Co	59	*Silicon*	Si	28
Copper (Cuprum)	Cu	63	Silver (Argentum)	Ag	107.7
Didymium	Di	142.3	Sodium (Natrium)	Na	23
Erbium	E	166	Strontium	Sr	87.5
Fluorine	F	19	*Sulphur*	S	32
Gallium	Ga	69	Tantalum	Ta	182
Gold (Aurum)	Au	196.2	*Tellurium*	Te	128
Hydrogen	H	1	Terbium	Tb	—
Indium	In	113.6	Thallium	Tl	204
Iodine	I	127	Thorium	Th	232
Iridium	Ir	193	Tin (Stannum)	Sn	118
Iron (Ferrum)	Fe	56	Titanium	Ti	48
Lanthanum	La	138.2	Tungsten (Wolfram)	W	184
Lead (Plumbum)	Pb	207	Uranium	U	239.8
Lithium	Li	7	Vanadium	V	51.5
Magnesium	Mg	24	Ytterbium	Yb	173
Manganese	Mn	55	Yttrium	Y	89
Mercury (Hydrargyrum)	Hg	200	Zinc	Zn	65
			Zirconium	Zr	90
Molybdenum	Mo	96			

NOTE.—The names of metals are printed in Roman, non-metals in *italics*.

MORE IMPORTANT CHEMICAL COMPOUNDS.

Acid, acetic .... $CH_3CO.OH$.

" boric .... $B(OH)_3$.

" hydrochloric (muriatic)....................... HCl.

" hydrocyanic (prussic).......................... HCN.

" nitric.... .... HNO_3.

" oxalic............................... .... $(CO.OH)_2$.

" salicylic.............. ....$C_6H_4(OH)CO.OH$.

" sulphuric .... H_2SO_4.

" tartaric............................... ......$C_2H_4O_2(CO.OH)_2$.

Alcohol, amyl (fusel oil)......... .... $C_5H_{11}OH$.

" ethyl (common alcohol)..................... C_2H_5OH.

" methyl (wood alcohol)................,........ CH_3OH.

Alum.. ....$Al_2 3SO_4,K_2SO_4 + 24H_2O$.

Ammonium hydroxide (ammonia)....... .... NH_4OH.

" carbonate............................... $(NH_4)_2CO_3$.

" chloride............................ ...... NH_4Cl.

" nitrate .... NH_4NO_3.

Aniline. .... $C_6H_5NH_2$.

Arsenic trisulphide (orpiment)....................... As_2S_3.

Arsenious oxide (arsenic)................... .... As_2O_3.

Barium chloride...................................... $BaCl_2$.

" sulphate.................................... $BaSO_4$.

Benzene (benzol)..................................... C_6H_6.

Cadmium iodide....................................... CdI_2.

" sulphide (cadmium yellow)............... CdS.

Calcium carbonate........... .... $CaCO_3$.

" chloride.................................... $CaCl_2$.

" oxide (lime)..... .... CaO.

" sulphate (gypsum)................... ....$CaSO_4,2H_2O$.

Camphor..$C_{10}H_{16}O$.
Carbon dioxide (carbonic acid)................CO_2.
    "   disulphide........................CS_2.
    "   monoxide...........................CO.
Chloral hydrate...................................CCl_3COH,H_2O.
Chloroform............... ....$CHCl_3$.
Cobalt nitrate.....................................$Co2NO_3$.
Copper sulphate (blue vitriol)..................$CuSO_4,5H_2O$.
Ether (sulphuric ether).........................$(C_2H_5)_2O$.
Ferric chloride...................................Fe_2Cl_6.
Ferrous sulphate (green vitriol)................$FeSO_4,7H_2O$.
    "   sulphide.............................FeS.
Glycerin...$C_3H_5(OH)_3$.
Hydrogen sulphide (sulphuretted hydrogen).........H_2S.
Lead acetate (sugar of lead)....................$Pb(CH_3CO.O)_2$.
    " chloride ....$PbCl_2$.
    " chromate (chrome-yellow)....................$PbCrO_4$.
    " oxide (litharge)....................PbO.
    "   " (red lead)................Pb_3O_4.
Magnesium chloride.................................$MgCl_2$.
    "   oxide (calcined magnesia)...........MgO.
    "   sulphate (Epsom salts)..................$MgSO_4,7H_2O$.
Manganese dioxide (black oxide of Mn).............MnO_2.
    "   sulphate...................... $MnSO_4$
Mercuric chloride (corrosive sublimate)..............$HgCl_2$.
    "   oxide (red oxide of Hg)..................HgO.
    "   sulphide (vermilion)....................HgS.
Mercurous chloride (calomel).......................$HgCl$.
Nitro-benzene.....................................$C_6H_5NO_2$.
Potassium antimonyl tartrate (tartar emetic).......$K(SbO)C_4H_4O_6,\frac{1}{2}H_2O$.
    "   carbonate (pearlash)....................K_2CO_3.
    "   bicarbonate (saleratus)....................$KHCO_3$.
    "   chlorate.................................$KClO_3$.
    "   chromate................................K_2CrO_4.
    "   bichromate.............................$K_2Cr_2O_7$.
    "   ferricyanide ....K_3FeCy_6.
    "   ferrocyanide (prussiate of potash).......K_4FeCy_6.
    "   hydroxide (caustic potash)..............KOH.
    "   iodide..................................KI.

Potassium nitrate (niter, saltpeter)..................KNO_3.

"   permanganate................ ....$KMnO_4$.

"   bitartrate (cream of tartar)..............$KHC_4H_4O_6$.

Silver nitrate (lunar caustic)........ ....$AgNO_3$.

Sodium pyroborate (borax)............. ....$Na_2B_4O_7,10H_2O$.

"   carbonate (soda).............................Na_2CO_3.

"   chloride (common salt).......................$NaCl$.

"   hydroxide (caustic soda)....................$NaOH$.

"   sulphate (Glauber's salt)....................$Na_2SO_4,10H_2O$.

"   potassium tartrate (Rochelle salt)..........$NaKC_4H_4O_n,4H_2O$.

Stannic sulphide (mosaic gold).......................SnS_2.

Strontium nitrate....................................$Sr2NO_3$

Sugar, cane ....$C_{12}H_{22}O_{11}$.

"   grape (glucose)............... ....$C_6H_{12}O_6$.

Turpentine...................................... ... .$C_{10}H_{16}$. .

Zinc sulphate...................................... ..$ZnSO_4$.

DIRECTIONS FOR EXPERIMENTS.

THE following simple suggestions will enable any student to perform all the experiments mentioned in this work. Many easy illustrations are also given in addition to those named in the text. Carefully compare them with those contained in the body of the book. The full-faced figures refer to the pages of the book, and the light-faced figures to the number of the experiment.

INORGANIC CHEMISTRY.

**I. *Indestructibility of Matter.*—Two invisible substances are formed by the burning candle, as may be proved by the following experiments:

1. Hold a cool, dry tumbler, or glass flask, over a candle-flame. The glass will be immediately covered with a fine dew from the steam produced by the burning candle.

2. Lower a lighted candle-end, fixed on a handle of wire, into a clean glass bottle, and loosely cover the mouth of the bottle with a piece of paper. After the flame has gone out, remove the candle; pour a little clear lime-water (see p. 139) into the bottle, and shake. The lime-water becomes milky, while in another bottle in which no candle has been burned it will remain clear (prove this). This shows that the burning candle has formed an invisible substance, which, unlike air, can render lime-water turbid.

Weight and Impenetrability of Gases.—3. Weigh a flask or bottle full of air, and again after the air has been exhausted by means of an air-pump. While on the balance, allow the air to enter again.

4. Balance a bottle full of air, and then fill it with carbon dioxide (see p. 66).

5. The impenetrability of air may be shown by plunging a tumbler, mouth downward, into water. The water rises but a short distance into the tumbler.

**3. *Chemical Action of Light.*—6. Dissolve a little salt and a little silver nitrate in separate portions of water. On mixing the solutions, a white precipitate is formed, which, on exposure to sunlight, turns dark.

Effect of Solution in Promoting Chemical Action.—7. The sodium carbonate and tartaric acid may be intimately mixed by grinding them together in a mortar. No chemical action will take place till water is added.

**II. *Oxygen.*—8. Heat 4-5 grams of red oxide of mercury in a hard glass tube, sealed at one end. Mercury is deposited on the sides of the tube beyond the flame, and a gas is given off which may be proved to be O by its kindling a match on which a spark has been left. By continuing the experiment, all of the HgO can be converted into Hg and O.

9. Put into a dry test-tube a few grams of pure potassium chlorate, and heat cautiously. The test-tube may be supported by a strip of thick paper twisted around it at the top. Move the tube to and fro through the flame at first, until it becomes fully heated; keep the tube inclined, and not perpendicular, letting the flame strike the side rather than the bottom. Hold the thumb lightly over the mouth of the tube. The salt melts quietly, and then begins to decompose, with the appearance of boiling. That O is given off is proved as in Ex. 8. When no more gas is evolved, allow the salt to cool, shaking it gently to prevent its attaching itself to the tube. The residue is no longer a *chlorate*, and gives out no yellow gas if moistened with H_2SO_4, but will yield a white precipitate with $AgNO_3$, which shows it to be a *chloride*.

10. Most of the following experiments may be performed in test-tubes as above, but, when it is desirable to make a larger quantity of O, one ounce of potassium chlorate is *very carefully* pulverized, and mixed with half that quantity of black oxide of manganese.* Be careful not to grind them together. The mixing is effected by placing them both on sheets of clean paper, and pouring them back and forth from one sheet to the other until the mixture has a uniform gray color. Place the mixture in a flask; fit a cork to the nozzle; then withdraw the cork, and with a round file bore a hole through it just large enough to admit a glass tube bent† as shown in Fig. 1. Return the cork and tube, arrange the apparatus as shown in the figure, and apply the heat. This must be done very cautiously at first, holding the lamp in the hand, and moving it around so that the flame may strike all the lower part of the flask, and thus expand it uniformly. Be careful, also, that no draft of cold air strikes against the heated glass. The first few bubbles of gas will consist mainly of the air contained in the flask, and should not be caught. When the gas begins to pass over freely, diminish the heat. *When the gas ceases, remove the stopper from the flask, or lift the end of the tube out of the water;* otherwise, as the flask cools, *the water in the tub will rush back into the flask, and break it.* When the retort is nearly cool, pour in some warm water to dissolve the residuum, which may then be poured out, and the flask dried for future use.

Instead of bending the glass tubing, it may be cut into short lengths, and the pieces joined by bits of rubber tubing. The advantage of this is that the flexible joints are not liable to break, and the apparatus may be more easily moved. Where a large quantity of O is to be made, a copper

---

* In order to test the purity of the materials, and thus avoid any danger of an explosion, it is well, previous to putting the mixture in the flask, to place a little in an iron spoon, and heat it over the lamp. If the gas pass off quietly, no danger need be apprehended.

† A glass tube may be bent at any point by softening that part in the flame of an ordinary gas-burner. Practice alone will give the required expertness. The following points should be observed: 1. Keep the tube slowly turning between the fingers, so that it may be equally heated on all sides; 2. Do not twist or pull the tube while heating; 3. Do not bend it until very soft; if not hot enough, the elbow will be flattened. Blowing gently into the tube at the moment of bending also prevents flattening.

retort and rubber tubing will be found cheap and convenient. No especial care is then needed in managing the heat.*—In place of the pneumatic tub, a pail or a tin pan may be used, letting the bottle rest on a shelf, as in Fig. 8, or on a couple of bricks.—The bottles for collecting the gas may be the regular "deflagrating jar " of the chemist, or the common "packing bottle " of the druggist. They are to be sunk in the water of the pneumatic tub, and filled; then inverted, and lifted upon the shelf, carefully keeping the lower edge of the bottle under the water. The bottles may also be filled from a pitcher, then closed with the hand or a plate, and quickly inverted and placed on the shelf in the tub or pan ready for use. As soon as a bottle is filled with gas, a plate may be slipped under the mouth, and thus, leaving enough water in the plate to cover the lower edge, be set aside, as in Fig. 1. Gas may be passed from one jar to another in the manner shown in Fig. 18.—While the gas is being collected, the water from the bottles which are filling may cause the tub to overrun; to prevent this, arrange a siphon to carry off the water into a pail below the table.—When a jar of gas is wanted for use, slip a plate under the mouth, or simply close it with the hand, and, lifting the jar out, carry it to the table, and place it mouth upward. Uncover only when the experiment is ready to be performed, as the gas will slowly diffuse.

¶3.—11. The experiment with the candle may be very strikingly performed by filling a common fruit-jar with O, and another with N. The covers may be loosely laid on top, and the lighted candle passed quickly from one to the other, as mentioned in note on page 30. The candle may be simply stuck on the end of a bent wire, as in Fig. 15, but it is much neater to have the tinsmith fit a little cup for its reception.

¶4.—12. If brimstone be used in the experiment with S, and it fails to light readily, pour upon it a few drops of alcohol, and then ignite.

13. Worn-out watch-springs can be obtained gratis of any jeweler, and may be easily straightened by slightly heating, and then drawing them between the fingers. If the end of each spring be strongly heated, and then pounded with a hammer on any smooth, hard surface, the temper may be drawn, and the edge sharpened. Make a slit with a knife in the side of a match, into which insert the edge of the spring. Take a piece of zinc or tin large enough to cover the mouth of the jar containing the O, and make a hole through it with a nail. Pass the other end of the spring through this hole, and then through a thin cork. The spring is now ready for burning. The metal cover will prevent the flame from coming out of the jar and burning one's hand, and the cork will hold the spring in its place.† When

---

* If, during the operation, the gas suddenly ceases to come off, remove the flame, and ascertain whether the delivery-tube is not choked up, which would result in a very violent explosion of the retort.

† It is well to obtain several pieces of thin board or shingles, about six inches square, bore a small hole in the center, and insert a match end or small plug. These may be used as covers in most experiments where a deflagrating spoon is to be employed. The handle of the spoon is passed through this hole, and held in place by the small plug.

the match is ignited, and then lowered into the jar of O, the spring should not reach more than half-way to the bottom, and should be pushed down as it burns. A cheap packing bottle should be used, as the glass is frequently broken by the melted globules of iron. Do not fill it quite full of gas, as then, on inverting, a little water will be left at the bottom, or some fine sand may be thrown into the jar before the experiment. The illustration may be repeated with a coil of fine iron wire. The springs from an old hoop-skirt burn nicely in O.

14. When S, P, charcoal, a wax candle, Na, and other substances are to be burned in O, they may be supported on the end of pieces of glass tube bent like the letter J, and left open at the shorter end only. For this purpose, covers must be provided with holes large enough for the glass tube to pass through. Or a "deflagrating spoon" may be readily extemporized to contain the phosphorus. Hollow a small piece of chalk, and attach a wire to it, which may then be secured to a metal top, as in the case of the watch-spring. This need not be pushed down into the jar as the burning progresses. Be careful to cut the phosphorus under water, to dry it carefully with blotting-paper, and not to handle it. The fumes are very disagreeable, and should not be inhaled or allowed to escape into the room. They soon dissolve when shaken with a little water.

15.—15. In burning bark charcoal in O, force the gas into the bottle through a bit of rubber tubing at the mouth. By placing this at one side, the gas is given a rotary motion, and the sparks of ignited charcoal will drive around the bottle in a beautiful maelstrom of fire. The gas may be forced in, from a rubber bag, and this striking effect easily produced.

16. Arrange a receiver upon the bed-plate of the air-pump so that O may be admitted from a gas-bag by turning a stop-cock. Put under the receiver an ignited tallow candle, with a big wick. Exhaust the air until the flame goes out, and there is left only a coal of fire. Admit the air, and it will have no effect to restore the blaze. Force in some O quickly, and the coal will burst instantly into a brilliant white light, brighter than at the first.

17. The oxidation of one solid by means of nascent O, liberated from another solid, can be shown as follows: Heat about five grams of potassium nitrate (saltpeter) in a test-tube until it melts quietly. Remove the lamp, and throw in pieces of S as large as peas, when they burn with an intensely bright flame. The heat is often sufficient to melt the glass, and the precaution should be taken to hold it over an iron plate or sand-bath. Or melt a quarter of a pound of saltpeter in an evaporating dish; an ordinary tin cup will answer. Put it on some burning coals in a draught to carry off the fumes. Plunge into the liquid a piece of bark-charcoal, strongly ignited. The oxygen of the saltpeter will support the combustion, and the charcoal will deflagrate in a rushing volcano of scintillations.

23. *Ozone.*—18. Scrape off the white coating of a stick of phosphorus under water. Place it in a wide-mouth liter-bottle full of air, with about a teaspoonful of water at the bottom. Close the mouth of the bottle with a glass plate, and expose the whole for an hour or two to a temperature of

15° or 20° C. Then invert the neck of the bottle in water, and allow the phosphorus to fall out. Replace the glass plate, and withdraw the bottle and its contents from the water. The phosphorus in this experiment undergoes a slow oxidation, during which a little ozone is formed, and is left mixed with the air; but the ozone will be again destroyed if it is left too long with the phosphorus.

19. Put in an evaporating dish a little starch; cover it with water in which a few crystals of potassium iodide have been dissolved, and heat. Stir the liquid, to prevent lumps. When cooked, immerse in the paste slips of white blotting or clean writing-paper, and hang them up to dry. They must be moistened when used.

20. Let some ozone pass into a clean bottle containing a little pure mercury. Shake the whole very carefully. The metal will change so as to act like an amalgam of tin and mercury, and will form a mirror on the sides of the bottle.

28. *Nitrogen.*—21. The phosphorus will, without the aid of heat, gradually remove the O from the air, forming phosphorus trioxide (P_2O_3), which will be dissolved by the water, and in a day or two the gas which is left will be nearly pure N. To show the proportion of O and of N in common air: Take a long glass tube; seal one end air-tight; with a camel's-hair brush and black paint mark upon the outside the division into fifths, and introduce a bit of phosphorus on the end of a long wire. Place the tube upright, with its open end under water. The water will gradually rise in the tube until it fills one fifth of the tube.

31.—22. For making HNO_3, take equal weights of sodium or potassium nitrate and strong sulphuric acid. The fumes may be caught in an evolution-flask, which is kept cool by water. When the retort is partially cooled, at the conclusion of the process, pour in a little warm water, to dissolve the potassium sulphate, otherwise the retort may break by the crystallization of the salt.

23. To show the effect of HNO_3 upon the metals, procure bits of tin and copper from the tinsmith. Take six wine-glasses, and place them in a row upon ordinary soup-plates containing a little water. Cover each with a beaker-glass or bell-jar. In one put a strip of copper, in another a little mercury, in another a piece of pure tin (not tinned iron), in another a strip of zinc, in another a new iron nail, in the last a bit of platinum wire or foil. Pour strong nitric acid upon each, and cover *immediately*. The copper, mercury, and zinc dissolve with a violent evolution of gas. The tin is oxidized to a white powder, while the iron and platinum are unaffected. Touch the iron with a piece of zinc, and it begins to dissolve. Put another new nail in *dilute* nitric acid, and it is rapidly dissolved.

24. Mix slowly together one ounce oil of vitriol and two ounces of the strongest nitric acid. When cold, dip paper into the mixture, and quickly wash with cold water and dry. The paper will burn with a flash like gunpowder. To avoid getting the acid on the hands, use glass tubes or rods for taking the paper out of the acid. Cotton treated in the same way becomes soluble in a mixture of alcohol and ether, and is used by photographers in making collodion.

33.—25. A special apparatus is necessary both for preparing and in-haling nitrous oxide safely. This consists of a glass retort—as shown in the cut—a wash-bottle, and, in addition, a gas-bag of from twenty to fifty gallons capacity for storing the gas, and a smaller bag of from three to five gallons, with a wide, wooden mouth-piece for inhalation. It is well to pass the gas through a large wash-bottle half full of caustic potash solution, and a second half full of H_2O, as shown in Fig. 13, thence by a rubber tube directly into the large gas-bag. The utmost care should be taken both in preparing and administering this gas, as other oxides of nitrogen are liable to be present, especially if too high a heat is used. Before preparing the gas, pour into the bag a couple of gallons of H_2O, by standing over which it will be purified in a few hours. When about to administer the gas, let the subject grasp his nose firmly between his thumb and forefinger; then, inserting the wooden mouth-piece, be careful that he does not inhale any of the external air, but takes full, deep breaths in and out of the gas-bag. Watch the eye of the subject, and notice the influence of the gas. Great care is necessary, and no one should ever inhale the gas who is not in good health, who is troubled with a rush of blood to the head, any lung or heart disease, or is of a plethoric habit. N_2O should never be admin-istered except when prepared and given by an experienced person.

Repeat experiments 11, 12, and 13 with this gas.

26. Half fill a test-tube with gas, over water. Close the tube under water firmly with the thumb, and then agitate the water and gas together. On removing the thumb under water, a considerable rush of water into the tube will occur, as the gas is soluble in about its own volume of cold water. By this circumstance the gas is easily distinguished from O.

34.—27. When a jar is filled with the NO, it may be lifted out of the H_2O and inverted, when the NO_2 will pass off in red clouds. If the jar be left in the cistern, and one edge be lifted so as to admit a bubble of air, red fumes will fill the jar. By standing a moment, the water will absorb the red vapor. The process may be repeated several times with the remaining gas. The variation of this experiment, described in the note on page 34, will be found very interesting. The change of color produced by mixing nitric oxide with any gas containing free O, often affords a convenient means of detecting small quantities of O when present in admixture with other gases, such, for instance, as coal-gas. Hence NO may be used to dis-tinguish between O and N_2O.

28. Into a large jar inverted over water, introduce a measured quantity of NO, and exactly half as much pure O. The two combine to form NO_2, which is soon dissolved in the water, and disappears. This illustrates Gay Lussac's law that gases combine in simple proportions by volume.

35.—Ammonia is so much lighter than air, that it may be conveniently collected by upward displacement, as shown in Fig. 10.

36.—29. If the bottle used for collecting the NH_3 be removed, and the flame of a Bunsen burner be applied to the jet of issuing gas, the NH_3 will not burn, but the gas-flame will be tinged with a pale yellow color. To show the burning of NH_3 in O, lead O into a wide-mouthed flask containing strong aqua ammonia (Fig. 12). On gently heating, the ammonia, mixed with O, will come off, and may be lighted at the mouth of the flask.

39. *Hydrogen.*—30. For preparing H, the apparatus shown in Fig. 13 is very convenient. The wash-bottle, *d*, is necessary only when it is desired to purify the gas for inhaling. A common junk-bottle, fitted with a cork and a glass tube, will answer for all ordinary experiments, but a "hydrogen generator," as described below, is much more satisfactory. The Zn for making H should be granulated.* Water may be poured into the flask until the lower end of the funnel is covered, before adding the acid. The flow of gas may be regulated by additions of acid, as may be wanted. One part of acid to ten or twelve parts of water will liberate the gas rapidly. If too much H_2SO_4 be added, the liquid is apt to froth over.

A constant hydrogen generator can be readily made by taking two bottles with tubulature near the bottom, such as are sold by druggists for the "nasal douche," and connecting them with a strong rubber tube. One is fitted with a cork and delivery tube provided with a stop-cock. In this bottle is placed a layer of pebbles or broken glass, and upon this a quantity of zinc scraps. In the other bottle is dilute H_2SO_4. On opening the stop-cock, the acid comes in contact with the zinc, and H is evolved. As soon as the stop-cock is closed, the pressure of the gas drives the acid back into the other bottle. The same kind of apparatus may be employed for generating CO_2 or H_2S.

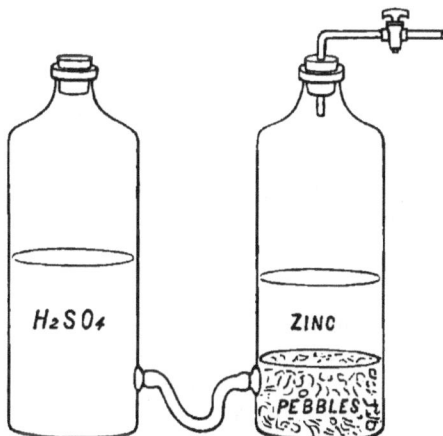

A hydrogen generator, similar in principle to the Döbereiner lamp (Fig. 19), can be made by cutting off the bottom of a tall and narrow bottle, filling it with zinc scraps, and closing at the lower end with a perforated rubber cork, and at the upper end with a perforated cork carrying a brass or glass tube and stop-cock. Place it upright in a jar of dilute sulphuric acid.

In experimenting with H, great care must be used not to ignite the jet of gas until all the common air has passed out of the flask; otherwise a severe explosion will ensue.† It is a safe precaution to test the gas by

_____

* This is easily done by melting the Zn in an iron ladle, and pouring the metal slowly from a little height into a basin of water.

† Always wrap a cloth around the H generator when you ignite the gas, as an additional precaution.

passing it in bubbles up through H_2O, and igniting them at the surface, the force of the combustion will indicate if there be any danger. H must not be kept in bags for any great length of time, as the air will gradually force itself in, and the gas will partly pass out, thus forming an explosive mixture which it is dangerous to ignite.

31. The gases may be mixed in the following manner: Fit a good cork into the neck of a large jar, and pass through it a tube five centimeters long. Bind a short piece of rubber tubing firmly to the tube, and close this elastic tube with a small pinch-cock.* Fill the jar with water over the pneumatic tub. Fill a small jar, which will hold about half a liter, with O, and transfer it, as shown in Fig. 18, to the large jar. Fill the same jar with H, and transfer it to the large jar. Repeat the operation with the H, so as to obtain in the larger jar a mixture of half a liter of O and one liter of H. Having previously softened a thin bladder by soaking it in water, tie into the neck of it a glass tube five centimeters long; then adjust to the projecting portion a piece of rubber tubing provided with another pinch-cock. Press the air out of the bladder; connect the two pieces of rubber tube by means of a short piece of glass tubing; depress the jar in the pneumatic tub, and then open each pinch-cock. The gas will now pass into the bladder,—if it does not, press the jar deeper into the water. Close both pinch-cocks, and remove the bladder. Now place the end of the tube attached to the bladder under some soap-suds, and blow a mass of soap-bubbles by squeezing the bladder. *Remove the bladder to a distance*, and then apply a light to the bubbles. A loud explosion will immediately follow.

32. A clay tobacco-pipe may be attached to the gas-bag by means of a bit of rubber tubing. Dip the pipe-bowl into the soap-suds, and, lifting it out, blow a bubble with the mixed gases, and then detach it by a quick motion. When the gas-bag is removed, ignite the bubble, which will explode sharply. If bubbles be blown with H alone, they will rapidly rise, and, if out-of-doors, will float to a great distance.†

33. H is the lightest known substance. Fill two bottles with the gas, suspend one inverted from the ring of the retort-holder, and place the other right side up on the table. In a few minutes the upright cylinder will be found to contain little or no H, while the inverted bottle is nearly full.

34. Suspend an *inverted* beaker glass from the end of the scale beam (Fig. 27). Balance it carefully; then fill it with H gas. It will be found much lighter than before.

35. Take a small porous cup, such as is used for electrical batteries. Fit a cork to it, and pass a long glass tube through the cork. Cover the cork

* Small pinch-cocks are sold for this purpose. They are cheaper than stop-cocks, and answer every purpose. In lieu of these, common spring clothes-pins may be used.

† If one has a large rubber gas-bag, with stop-cock and rubber tubing, and a glass receiver fitted with a stop-cock on top, these may be attached, and the gases measured in the receiver, and then passed directly into the bag. Such apparatus, though convenient, is not necessary to illustrate the properties of the gases.

with plaster of·Paris. Place the tube upright, with its lower end dipping into a colored liquid (CuSO, + NH,OH). Hold a jar of H over the porous cup. The H enters the cup, driving the air out of the lower end of the tube. Remove the jar, and the liquid will rise several inches in the tube. (See "Physics," p. 50.)

36. A substitute for spongy platinum in the experiments with hydrogen gas: Make a cylinder of pumice-stone, three eighths of an inch in diameter. With a fine saw, cut it into disks about one twentieth of an inch thick. Soak these for some time in a strong solution of bichloride of platinum in alcohol, and then as long in an alcoholic solution of sal-ammoniac. After being once thoroughly ignited, these disks will inflame a jet of hydrogen.

45.—37. The analysis of water can be readily performed as follows: Take a wide bottle (the height is unimportant), and cut it off about two and one half inches below the neck, by making a scratch with a three-cornered file, and then applying near the scratch a very hot piece of wire or glass, or a small blow-pipe flame. The crack will follow the flame slowly around the bottle. Take two strips of platinum foil, and put one on each side of a well-fitting cork, so that one end extends into the bottle, the other outside of the neck. Support the apparatus inverted upon a ring of the retort stand, and fill nearly full of water acidulated with H,SO,. Connect the strips of Pt with the poles of a battery of two bichromate or Bunsen cells. Bubbles of gas at once appear, and can be collected in inverted test-tubes, and tested. Instead of a bottle-neck, a broken funnel may be employed. (See "Physics," p. 237.)

38. The synthesis of water may be shown, and also its composition by weight, by passing dry H over dry CuO. The CuO is placed in a bulb-tube

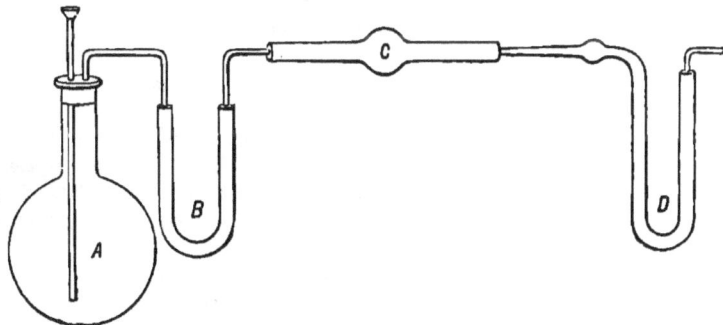

of hard glass C, and weighed. The tube D, filled with pieces of fused calcium chloride, is also weighed. The H generated in A is dried by CaCl, at B, and passes over the CuO at C. When the air has *all been expelled*, heat the CuO until it has a bright red color. Allow the H to pass through until the tube C is cold. Take the apparatus apart, and weigh the tube C; it will have decreased in weight by a quantity equal to that of the O expelled. Weigh the tube D. It has increased by a quantity equal to that of the water formed. If C has lost 16 grams in weight, D will have increased 18 grams in weight, showing that 16 parts of O will form 18 parts of water, by taking up 2 parts of H.

39. Burning H in O. Attach a $CaCl_2$ drying tube to a hydrogen generator, and to this a glass tube bent twice at right angles, and then turned up at the end, as shown in the figure. Take a small piece of Pt foil, and roll it around a darning-needle so as to form a small tube. Soften the end of the glass tube, and slip this little tube into it while hot, then hold them in the flame until the glass settles down against the Pt on all sides. When the air has all been expelled from the generator, throw a towel over it, and ignite the H, then bring it into a broad jar of O. The heat is very intense, hence the need of a Pt tip.

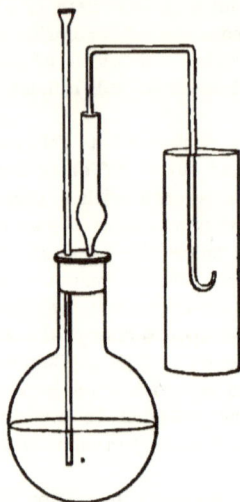

40. Burning O in H. To show the reverse experiment of burning O in H or illuminating gas, take a large lamp chimney, and fit a cork in each end. In the upper cork, insert a glass tube drawn out to a jet D. In the lower end, insert a bent glass tube A, and a metal tube C, made by rolling up a strip of sheet iron or brass. Fit a cork to the metal tube, and pass a glass tube with a Pt tip on it through this cork (B). Attach the tube A to an H generator, and when the whole apparatus is *full* of H, ignite it at D and C. Connect B with a gasometer or bladder of O, then quickly insert the cork into the opening C, so as to extinguish the flame there. Let the O pass in *slowly*. The O will be seen to burn in an atmosphere of H. Cut off the supply of O, and extinguish all flames before removing the supply of H, to avoid an explosion.

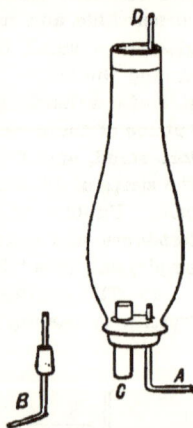

48.—41. Take some fresh crystals of sodium sulphate; let them lie exposed on a piece of blotting-paper for two or three days. They will gradually lose their water and crumble down, or *effloresce* into a white powder. Common washing soda will do the same.

42. Take two four-ounce bottles, and put in each a teaspoonful of white sugar. Fill one bottle nearly full of pure rain or distilled water, the other with impure water. Cork them, and let them stand a few days; in one bottle will be seen a fungoid growth, resembling fuzz or lint; the water in the other bottle remains clear. (Note, p. 194.) The sporules or germs that

fall into the water find suitable nutriment for their development in one case, but not in the other. *Any well or spring water in which this change takes place is unfit to drink.* (Page 48, note.)

43. Select a thin, porcelain dish which will hold 60 or 80 cub. cm.; place it in one pan of the balance, and trim a piece of lead until, when placed in the other scale-pan, it will counterpoise the dish. Measure a quarter of a liter of spring-water, and pour some of it into the weighed dish; place it over a very small gas-flame, so as to evaporate the H_2O gently, without allowing it to boil; add the rest of the H_2O from time to time until it has completely evaporated. Dry the salts thus obtained, and weigh what is left as accurately as you can. By multiplying this quantity by 4, you will obtain the amount of soluble solid substances per liter which that particular specimen of water contained. This is the basis of the plan which, with many additional precautions, is adopted for determining the quantity of salts in the process of analyzing waters to be used for drinking or manufacturing purposes.

50.—44. The expansion of water in freezing may be shown thus: Fill a common round-shouldered bottle with ice-water; cork tightly, and place in a freezing mixture (broken ice and salt); or, if it is a cold winter's day, out-of-doors. The ice in crystallizing will either break the bottle, or push out the cork, which will be found frozen on to the end of a stem of ice. If the bottle be broken, it will be seen that it is completely filled with solid ice.

55. Carbon.—Small paste-diamonds may be obtained of a jeweler, to illustrate the forms of cutting the diamond.

56.—45. Take a glass cylinder open at both ends, and suspend near the top an inverted funnel which fits nicely into the cylinder. Place a bit of camphor burning on a small dish below it, so that the smoke passes up into the funnel. When a considerable quantity of lamp-black has formed on the sides of the cylinder and funnel, extinguish the flame and remove the lamp. Then slowly lower the funnel, the edge of which will scrape the lamp-black from the sides of the vessel, in the same manner that it is done by manufacturers in making lamp-black for the market.

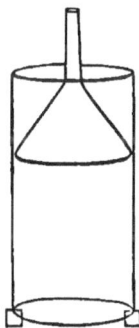

62.—46. Place a filtering-paper * in the glass funnel, and in it two ounces of bone-black or finely powdered charcoal. Filter through it water colored with ink, litmus, or any other impurities. In pouring the liquid into the filter, hold a glass rod against the edge of the pouring vessel, so as to direct the stream into the funnel. The funnel may be placed in the nozzle of a bottle, but must not fit closely. A bit of wood or a thread inserted between the stem of

---

* In order to prepare this filter, fold a square of filter-paper, as shown in the figure on next page, first into half, and then again into a quarter of its first size (*b*); cut off the edges in the direction of the dotted line shown in the left-hand figure (*a*), open out the folded paper (*c*), and drop it into a funnel a little larger than the paper-cone.

a b c

the funnel and the nozzle will leave an opening sufficient for the egress of the air.

47. Slip a piece of freshly-burned charcoal under the edge of a long tube previously filled with dry ammonia gas,* and standing over Hg. The charcoal will quickly absorb the NH_3; the whole of the gas, if pure, will disappear, and the Hg will fill the tube.

48. Weigh a piece of freshly-burned charcoal as soon as it is cold; leave it exposed to the air for twenty-four hours, and weigh it again; it will be found to be heavier. Place the charcoal in a glass tube, and heat it over a lamp; moisture will be driven off, and will become condensed on the cold sides of the tube.

49. Shake up with a little powdered charcoal some stagnant water which has been kept till it smells offensively. In an hour it will have lost all its disagreeable odor.

50. Mix in a mortar twenty grams of litharge with forty grams of NaCl and one gram of powdered char oal; cover with a little more salt, and place the mixture in a small clay crucible; heat it to bright redness in the fire. When the mixture is melted, take the crucible out of the fire and let it cool. When quite cold, break the crucible, and a bead of Pb will be found at the bottom, under the melted salt, the C having taken the O from the PbO.

63.—51. Break some marble into small bits; place them carefully in the evolution-flask, and, inserting the cork and tube, pour in HCl slowly. The gas, on account of its weight, may be passed directly into a bottle or jar, and collected by downward displacement.

52. Lower a lighted candle into a jar of the gas, or, placing the candle in an empty jar, pour the gas into the jar, as if it were water.

53. Test the gas with moistened blue litmus-paper.

64.—54. In a pint of water place a piece of lime as large as an egg; let it stand over night; pour off the clear liquid; it is lime-water. Place a little in a tumbler and pass a current of CO_2 from the evolution-flask into the liquid. It becomes milky, and then, after a time, clear. If the clear liquid be boiled, CO_2 is driven off, and the lime is again precipitated. This is a

---

* The gas may be dried by passing it through a tube filled with pieces of lime.

good illustration of the formation of hard water, and of the way in which it produces an incrustation when boiled. (See p. 49.)

55. Repeat the experiment by breathing from the lungs through a tube into the lime-water. The last portions of air exhaled will be found to be most highly charged with CO_2.

Or, the formation of carbon dioxide in the lungs may be shown by an experiment of Faraday's. Fit an open-mouthed receiver with a cork, through which passes a small bit of glass tubing. The receiver is then placed in a basin of water. Expelling the air from his lungs, the experimenter inhales the air in the receiver through the tube. The water in the basin rises in the receiver, and shows how fast and when the air is exhausted. Breathing back the air from the lungs into the receiver again, the water is expelled, and the lighted taper will test the presence of the carbon dioxide. Before testing, the air may be breathed back and forward two or three times, until it becomes unpleasant. The rise and fall of the water in the jar is a pleasant and instructive addition to the experiment.

65.—56. Arrange little wax-tapers in a wooden or pasteboard trough, as on page 65. Light them, and then pour in at the top a bottle of carbon dioxide gas. If the proper slant is given to the trough, all the candles will be extinguished. (Fig. 26.)

57. Balance a beaker on a delicate pair of scales, or in any simple manner one's ingenuity may suggest. Empty into it a large jar of CO_2, and it will quickly descend. (Fig. 27.)

66.—58. Twist a wire around the neck of a small, wide-mouthed vial, to answer as a bucket. Lower it by the wire into a jar of CO_2, our ideal well foul with the gas. Raise it again, and test for the CO_2 by means of a lighted match. The bucket will be found full of the gas.

70.—59. Carefully heat in a flask fitted with a cork and gas-delivery tube (see Fig. 8), 5 grams of potassium ferrocyanide and 50 c.c. of concentrated H_2SO_4. CO will come off freely, and will burn with a bluish flame. Be careful not to inhale the gas.

60. Place a few crystals of oxalic acid $(C_2H_2O_4)$ in a test-tube, pour on enough oil of vitriol to cover them, and then heat gently. Both CO and CO_2 will be evolved. On applying a flame to the open end of the tube, the CO will take fire and burn with a blue flame. To separate the two gases, pass them through a solution of KOH, which will remove the CO_2, and the CO can then be collected over water.

71.—61. CH_4 may be made by heating in a test-tube 2 grams of sodium acetate, 8 grams of caustic soda, and 2 grams of powdered quicklime.

72.—62. Introduce into a retort which will hold a liter, 10 c.c. of alcohol and 60 c.c. of strong sulphuric acid. Heat the mixture, and collect the gas over water; continue the experiment until the mass blackens and swells up considerably. The product consists at first chiefly of olefiant gas, mixed with ether-vapor; but toward the end it becomes mingled with SO_2. Pass it through a solution of potash, using a wash-bottle as shown in Fig. 13, and then collect in the gas-bag. Fit a piece of glass tubing, drawn to a fine point at one end, to the stop-cock of the gas-bag, by means of a bit of the rubber tubing. On turning the stop-cock and forcing out the gas, it may be ignited, when it will burn with a clear white light.

63. Mix C_2H_4 with three times its bulk of O and explode in soap-bubbles. Great care must be taken not to let the light approach the gas-bag containing the mixture.

64. Get a bit of bituminous coal about the size of a walnut. Pound it small, almost into dust. Fill an ordinary tobacco-pipe (one with a long stem is preferable) *nearly* full of the pounded coal, packing it down closely with your thumb. On the top press a disk of metal or a copper coin, and cover with a layer of plaster of Paris or some tough clay, reduced to the consistency of putty by being tempered with a little water. Heat the bowl of the pipe strongly, and a combustible gas will come out of the stem, which should now be held in a nearly vertical position. When no more gas is given off, and the jet of flame goes out, remove the clay or plaster covering. The residue in the bowl is coke.

65. Place some bits of pine wood in a glass retort provided with a perforated cork and delivery-tube. Connect this with an empty wash-bottle. On heating the retort, gas is given off, tar collects in the wash-bottle and charcoal remains in the retort. The charcoal used for making gunpowder is prepared in this way in large iron cylinders.

74.—66. Fit a cork to a small test-tube. Take out the cork, and pass through it a bit of glass tubing drawn to a fine point at one end, so as to act as a gas-burner. Place in the tube about two grams of mercury cyanide; replace the cork, and heat over a spirit-lamp. The test-tube may be supported by a strip of thick paper twisted around it at the top. Move the tube to and fro through the flame at first, until it becomes fully heated; hold the tube inclined and not perpendicular; letting the flame strike the side rather than the bottom. When the gas begins to come off, it may be ignited.

67. To show the formation of potassium cyanide from a nitrogenous body: Drop into a perfectly dry test-tube a bit of nitrogenous substance and a small piece of freshly cut K or Na. Heat carefully until it melts and a flash of light is seen. When cold, break the tube, throw the fused mass into a clean test-tube, and make the test for KCy as given below.

68. To test for the presence of potassium cyanide: Add to a solution of the suspected substance a few drops of solution of sulphate of iron and a slight excess of potash. Shake the precipitate a few moments with air in the tube, and add an excess of hydrochloric acid, when a blue precipitate, or a decided blue or green color pervading the liquid, will indicate the presence of a cyanide. Prussian blue is produced by this process.

81—69. The compound blow-pipe with gasometers, as shown in Fig. 39, is a serviceable apparatus. If gas-bags are used, the one for H should be twice the size of the one for O. A board should be laid on each bag, upon which weights may be placed, when ready for use, so as to force out the gas steadily. Always ignite the H first, and then turn on the O slowly until the best effect is produced. All the metals burn in the blow-pipe flame with their characteristic colors. Narrow slips should be prepared for this purpose. A cup for holding the chalk is necessary to show the lime-light. A piece of hard lime, whittled to about the size of a pencil, may be held in the flame to illustrate the principle.

89.—70. To a small gas-jar fit a good cork, through which pass a test-tube as shown in Fig. 43. Place the jar in a large beaker glass or open-mouthed bottle, filled with spring water, which has been mixed with a fourth of its bulk of a solution of carbonic acid in water. Fill the tube with water, and place it in the neck of the jar, having introduced a few sprigs of mint or the leafy branches of any succulent plant; then expose for an hour or two in direct sunshine. Bubbles of gas will be seen studding the leaves; and on shaking the jar they will become detached, and will rise into the test-tube. After a time the cork and tube may be withdrawn, keeping the mouth of the tube beneath the surface of the water; then close it with the thumb, turn the tube mouth upward, and test the gas with a glowing splinter. The wood will burst into a blaze, showing that the gas consists mainly of O.

92. *Chlorine.*—71. Put in the generating flask (see Fig. 44) a mixture of equal parts of NaCl and MnO_2; insert the cork with its tubes, and pour in through the funnel tube sulphuric acid which has been diluted with an equal weight of water. On gently heating, the gas will come off abundantly, and may be collected by downward displacement in tall jars (see Ex. 51). The gas may be dried by passing it through a wash-bottle containing strong H_2SO_4.

93.—72. Plunge a lighted candle into the gas: it will burn feebly, with a red, smoky flame. (Fig. 45.)

73. Place a piece of dry phosphorus in a copper deflagrating spoon; introduce it into a bottle of Cl: the phosphorus will take fire and burn, while suffocating fumes of phosphoric chloride (PCl_5) are formed.

74. Dip a strip of blotting-paper into oil of turpentine; plunge it into a jar of Cl: it will immediately burst into flame, while a dense black smoke is given off. (Fig. 46.)

75. Powder some metallic Sb finely in a mortar, and sprinkle into a jar of Cl: it will take fire as it falls, giving out fumes of antimony chloride ($SbCl_3$), which are very irritating.

94.—76. Fill a flask with water and lead Cl into it until no more dissolves. Withdraw the tube through which the Cl has entered; fill the flask to its mouth with chlorine water; close it with a cork in which is a small, short glass tube; and support it in an inverted position, so that the end of the glass tube is below the surface of water contained in a beaker. Place in a window where the direct sunlight will fall upon it. A bubble of gas will soon appear at the top of the inverted flask, and will go on increasing in size from day to day until all the greenish color has disappeared from the water. Then transfer the gas to a test-tube (see Fig. 18) and test it with a spark on a splinter of wood. The gas is thus proved to be O. The water in the flask contains HCl, as may be shown by adding a few drops of silver nitrate solution (see p. 97).

77. Pour a little boiling water upon some chips of logwood, so as to

obtain a deep red liquid; add some of the solution of Cl, and the red color will be discharged.

95.—78. Write a few words with ordinary ink on a printed card and put it in moist Cl or in chlorine water. The writing will be bleached, while the printed words are unchanged.

79. Print the word PROTEUS on a large card, first with the iodide-of-potassium-starch solution (note, p. 101), and second with a solution of indigo. The former will be white and almost invisible, the latter blue. Then paint the words (using a camel's hair brush) with a solution of chlorine. The first line will turn blue and the second white, thus just reversing their color.

80. Bleach some colored calico by putting it into water with which a little bleaching powder has been mixed. The action is hastened by adding a little vinegar or a few drops of any dilute acid.

96.—81. Burn H in Cl, as shown in experiment of burning H in O. (Ex. 40.)

82. Wrap a soda-water bottle in a towel; fill it with water, and invert it in the pneumatic tub. Introduce a glass funnel into the neck, and having filled a jar of 100 cc. capacity with Cl, pass the gas into the bottle. Fill the same jar with H, and empty into the same bottle; withdraw the funnel, close the neck with the palm of the hand, lift the bottle out of the water-bath, give it a shake to mix the gases, and apply a light. A sharp explosion will immediately follow, and gaseous HCl be formed. Equal measures of H and Cl unite in this way, and the gas produced occupies the same bulk that its components did when separate. Sunlight will also cause the explosion of a mixture of Cl and H.

83. To prepare HCl put some common salt in the generating flask (Fig. 47), and pour through the funnel tube about twice its weight of strong H_2SO_4. A solution of the gas — ordinary hydrochloric acid — is obtained by leading it through water, as shown in the cut. The gas itself may be collected for experiment by downward displacement, as in the case of Cl.

The gas may also be obtained by gently heating concentrated hydrochloric acid. In this case it should be dried by leading it through strong H_2SO_4 in a wash-bottle, before collecting it for experiment.

84. A candle lowered into a jar of HCl gas is extinguished.

85. Fill two jars of the same size, the one with HCl gas, the other with NH_3. Bring them mouth to mouth and remove the glass plates by which they were closed. The two gases immediately combine, forming a white cloud of ammonium chloride.

86. Dilute a little HCl with six or eight times its bulk of water, and add caustic soda cautiously, until the liquid is neutral, and neither reddens blue litmus nor restores the blue to red litmus-paper. Pour the liquid into a basin, and evaporate it slowly; crystals of NaCl will be deposited in cubes.

87. Boil HCl in a test-tube with fragments of gold-leaf, or a bit of platinum wire; they will not be dissolved. Now add a drop or two of HNO_3; a yellow solution will be formed.

88. Fill a test-tube nearly full of pure rain or snow water, and add a drop or two of the nitrate of silver solution. A drop of HCl will cause a cloudy, white precipitate.*

100. *Bromine and Iodine.*—89. Pour a little strong H_2SO_4 onto a few crystals of potassium bromide in a test-tube. Heavy red vapor of Br will fill the tube.

90. Dissolve in a little water in a test-tube a crystal of potassium bromide; add a little chlorine water. Br is set free, and the solution becomes brownish in color; now add a few drops of chloroform, and shake thoroughly. When the chloroform settles to the bottom, it is colored yellow by the Br it has dissolved. If dilute starch paste is added instead of the chloroform, it also becomes yellow.

91. Repeat experiments 89 and 90 with potassium iodide instead of potassium bromide. Violet vapor of iodine will be given off in the first; the chloroform will acquire a beautiful violet color, and the starch paste will become dark blue.

103. *Sulphur.*—92. Melt a quantity of S, either the flowers or brimstone, in a test-tube. If heated carefully and uniformly, the liquid S is at first thin and amber-colored; then becomes dark, and so thick that at a certain point it will not run from the inverted test-tube; and finally, at a higher temperature, regains its fluidity and boils, giving off a deep red vapor, which readily burns. Pour the liquid S into water. It forms an elastic gum, which slowly hardens and becomes brittle.

93. Fill a clay crucible or a cup with brimstone, and melt it with a gentle heat. Set it aside to cool. When a crust has formed on top, break it, and pour out the liquid contents. If the cup be broken when cold, the interior will be found filled with long amber-colored transparent crystals of S, which after a time become opaque and yellow.

94. Pulverize some brimstone, and put in a test-tube and cover with CS_2. Put the tube in warm water until the S dissolves, then cork it tightly, or, better, pour the solution onto a watch-glass under a bell-jar, and let it stand until crystals separate. They have a different form from those obtained from fusion. The more slowly they separate, the larger they will be.

95. Put a piece of S in a test-tube, and above it some copper turnings or scraps. Heat the sulphur until it boils. The Cu burns brightly in the vapor of S. Fe and Pb will also burn in S vapor.

104.—96. Dissolve in a little water a crystal of potassium permanganate. On leading SO_2 into the solution, or adding a little sulphurous acid, the solution becomes colorless.

105.—97. Mix a little sulphurous acid and chlorine water. The presence of HCl and H_2SO_4 can be shown by the silver nitrate and barium chloride tests (pages 97 and 109).

108.—98. The method of making H_2SO_4 may be finely illustrated by

---

* Well water, which gives a considerable precipitate with $AgNO_3$, is usually unfit to drink, owing to sewage contamination.

means of the apparatus shown in Fig. 51. The large glass globe takes the place of the leaden chambers. Of the two flasks at the left, one contains strips of Cu and strong H_2SO_4, for the production of SO_2 (see p. 104); the other contains water, to furnish a supply of steam; NO is supplied from the generator at the right (see p. 33); air is introduced from a small bellows, or the mouth, through the long rubber tube, while the straight, upright glass tube acts as a chimney for the escape of waste gases. If the supplies of NO and air are cut off after the globe is filled with ruddy fumes, the gases filling the globe will soon become nearly or quite colorless, from the reduction of NO_2 to NO; on introducing a little air, the red color reappears.

99. Pour a little strong sulphuric acid into a test-tube. Place a splinter of wood in it; the wood will be blackened in a few minutes. Pour 1 cc. of strong H_2SO_4 into a tube containing 3 or 4 cc. of water; considerable heat will be felt to attend the mixture.* Take a little of this diluted acid, and, with a feather dipped into it, trace a few letters upon writing-paper. Hold the paper near the fire: the water will evaporate, leaving the acid behind; this will soon blacken the paper.

100. Mix 4 oz. H_2SO_4 and 1 oz. pounded ice or snow. Stir it with a test-tube containing a little ether; the ether soon boils.

101. Mix 1 oz. H_2SO_4 with 4 oz. snow or pounded ice, and stir it with a test-tube containing cold water; the water soon freezes. The vessel in which the experiment is performed usually freezes fast to the table, so that it is well to set in on a plate or small board.

102. Dissolve some sugar in a very little water, so as to form a thick syrup. Put it in a tall beaker, and pour on strong H_2SO_4 until it begins to swell up and blacken. The H_2SO_4 removes the H_2O from the sugar, leaving only C behind.

109.—103. Place in an evolution flask (A, Fig. 52) a few lumps of ferrous sulphide, and add some dilute H_2SO_4. Lead the H_2S gas through a solution of copper sulphate in B, one of tartar emetic in C, one of arsenic in D, one of zinc sulphate in E, and finally into water in the beaker. Precipitates—sulphides of the metals—will be produced, brownish-black in B, orange in C, yellow in D, and white in E. (See chapter on "Qualitative Analysis.")

110.—104. Place a few drops of the disulphide in each of four test-tubes. To one add a little powdered sulphur, to a second a few minute scales of iodine, to a third a fragment of phosphorus, and to a fourth a few drops of water. Notice the beautiful color produced by the iodine; the solution of the sulphur and the phosphorus; the insolubility of the liquid in water; and also its refractive power.

105. The experiment of burning P under water may be easily shown by throwing a piece of P into a glass of hot water. It melts, and looks like a thick oil under the water. A fine stream of O gas may then be passed through a glass tube down to the P, which burns brightly under water.

* In mixing H_2SO_4 and H_2O, always pour the acid *into* the water.

114. *Phosphorus.*—106. Dissolve 1 or 2 decigrams of phosphorus in 2 cc. of carbon disulphide in a test-tube; pour a little of the solution upon a piece of filtering-paper, and allow it to dry. The phosphorus will be left in a finely-divided form, and will set fire to the paper in a few minutes.

107. Place a bit of phosphorus in a solution of silver nitrate. In the course of a day or two, it will be covered with brilliant crystals of reduced silver. Repeat with CuSO, solution.

116.—108. To prepare hydrogen phosphide, place in the flask (*a*, Fig. 54) a strong solution of caustic potash and a few small pieces of phosphorus. The addition of a little ether will serve to expel the air, and prevent the danger of an explosion. Regular smoke-rings will be formed only in perfectly still air.

119. *Arsenic.*—109. Compounds of As, when heated on charcoal, give off a garlic odor.

110. Add a few drops of a solution of arsenic trioxide to 200 or 300 cc. of water, and then 3 or 4 cc. of HCl; place in the liquid two or three slips of bright copper foil, and boil the whole for a few minutes: the copper foil will become coated with a steel-gray film. Part of the Cu becomes dissolved and displaces the arsenic, which is thrown down on the undissolved portion. Pour off the water, dry the Cu on blotting-paper, and heat the foil in a tube, sealed at one end. The arsenic will sublime, condensing in minute octahedra on the cold sides of the tube. This is *Reinsch's test* for arsenic.

122. *Boron.*—111. Make a small round loop at the end of a thin platinum wire. Dip the hot loop into powdered borax, and heat the adhering salt until it melts to a clear, transparent bead. Moistened with a solution of cobalt nitrate and heated, the bead becomes blue; with manganese dioxide, or any manganese salt, the clear bead changes on heating to an amethyst color.

112. Add to a solution of boric acid (or borax and H_2SO_4), in a small dish, a little alcohol. Set fire to the alcohol, and, as it burns, stir the solution with a glass rod. The flame is tinged green.

113. A similar green coloration is obtained by bringing a little boric acid (or borax moistened with H_2SO_4) into the Bunsen flame on a platinum wire, which has first been dipped in glycerin.

Silicon.—114. Grind in the mortar 3 or 4 grams of fluor spar, and mix with an equal weight of powdered glass or sand. Introduce it into a flask previously fitted with a sound cork and a tube, as in the figure. Pour upon the mixture 30 grams of strong H_2SO_4, insert the cork and tube, and apply a gentle heat: a densely fuming gas is disengaged, consisting of silicic fluoride (SiF_4). This gas must not be inhaled, as it is very irritating. Pass it into a glass of H_2O, having sufficient Hg at the bottom to cover the mouth of the delivery-tube. Each bubble of gas, as it rises, is coated with a white film of hydrated silica, while a solution of hydrofluosilicic acid ($2HF,SiF_4$) is formed. The deposit of silica would clog the tube if it were not for the Hg; hence the tube must be kept dry, which is best accomplished by placing it in position in the Hg, then pouring water carefully into the glass on the Hg. Filter the solution, and preserve it as a test for Ba and K.

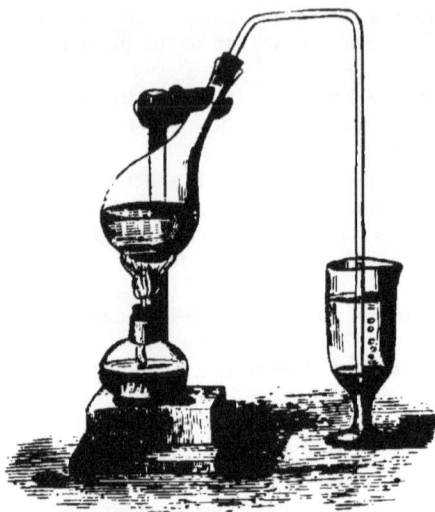

115. Grind a little glass to a fine powder in a mortar; place it on a piece of moistened red litmus-paper; sufficient alkali will be dissolved by the water to tinge the paper blue.

116. Mix a little fine sand with KNO_3 and Na_4CO_3, and heat strongly on a strip of Pt foil for five minutes. When cold, it will dissolve in H_2O.

117. Take some of the silica obtained in the experiment of making $2HF,SiF_4$, and put it in a strong solution of KOH, and boil. It will dissolve.*

118. Take four glass cylinders, five inches high, and pour into each about 1 oz. of ordinary water-glass and 4 oz. of water. Drop into one a few crystals of iron sulphate, into another some crystals of blue vitriol, into the third white vitriol, into the fourth a crystal of each. Let them stand quietly for twenty-four hours. In the first green fibers will be seen, resembling very closely a growing plant; blue and white ones will appear in the others. If closely corked, they can be kept for weeks.

110. Pour 1 oz. of water-glass into a capsule, and pour *on it* half as much H_2SO_4, taking care that the two do not mix. Pour immediately, but slowly, into a second capsule; the silica separates in long tubes resembling stalactites.

128. *Potassium.*—120. Place 30 grams of pearlash in a half-liter bottle, and dissolve it in 250 cc. of water. Shake 20 grams of quicklime with five or six times its bulk of water, and add the pasty mixture (about 120 cc. in bulk) to the boiling solution of pearlash. Agitate the mixture, and let it stand till it is clear. Pour off a portion of the liquid: it is a solution of caustic potash. Add to it some HCl: no effervescence will occur. Agitate a tablespoonful of olive-oil in a small vial with 3 or 4 cc. of the caustic solution diluted with ten times its bulk of water: a milky-looking liquid will be formed, which is the first stage in the making of soap.

---

* The infusorial silica, sold under the name of electro-silicon, will dissolve in KOH in the same manner. A basic silicate, known as "water-glass," or "soluble glass," is prepared in this way.

121 Burn some dry brushwood; collect the ash, wash it with five or six times its bulk of water, and filter. Test the solution with a piece of reddened litmus-paper, which will become blue. Evaporate the solution to dryness in a small porcelain dish. If the dry mass be left exposed to the air for a few hours, it will become moist. The potassium carbonate, of which it chiefly consists, attracts moisture rapidly, and deliquesces. To a portion of the salt, add a few drops of HCl: brisk effervescence occurs.

129.—122. Pulverize finely nitrate of potash and chloride of ammonium, five parts of each, and mix with sixteen parts of water. The temperature of the mixture will be reduced so low that, if a test-tube with a little water in it be used to stir it, the water in the tube will be converted into ice.

131.—123. Into a dilute solution of KOH, lead Cl gas until no more is absorbed. Evaporate the solution until crystals are formed. These consist of potassium chlorate, $KClO_3$. Separate them from the liquid, and dry by pressing between folds of filter paper.

124. To demonstrate the oxidizing power of $KClO_3$: Put into the mortar as much potassium chlorate as will lie upon the point of a knife-blade, and half as much sulphur. Cover the mortar with a sheet of writing-paper, having a hole cut in it just large enough for the handle of the pestle to pass through. When the two substances have become thoroughly mixed, grind heavily with the pestle, when rapid detonations will ensue. The paper will prevent loose particles from flying into the eyes. The same precaution should always be observed when pulverizing potassium chlorate. A better way is to purchase that salt in powder, or to make a hot saturated solution, and pour it out in thin films on panes of glass or old plates. It then forms very small crystals, which can be scraped off and dried for use. After using, clean out the mortar carefully for other experiments. The powder can be wrapped with paper into a hard pellet, and exploded on an anvil by a sharp blow from a hammer. Sometimes small bits of phosphorus are used instead of sulphur. Great care is then necessary, as the particles of burning phosphorus are apt to fly to some distance.

125. If four measures of a cold saturated solution of potassium bichromate be mixed with six of concentrated H_2SO_4, and the liquid be allowed to cool, chromic anhydride crystallizes in crimson needles, which may be drained and dried upon a brick.

The chromic anhydride is a powerful oxidizing agent, as may be shown by dropping onto some of the dry crystals a little strong alcohol. The alcohol will inflame, while the chromic anhydride turns green.

132. *Sodium.*—126. Take a small saucepan, and, having made a little pool of water upon a wooden stool, set the saucepan upon it; then throw in a handful of snow or powdered ice, and a handful of common salt; now stir with a stick, and the cold will freeze the saucepan to the stool, even before a large fire.

127. Fill a tall cylinder with a clear saturated solution of NaCl, and pass into it at the same time strong currents of NH_3 and CO_2 gases. A precipitate of $NaHCO_3$ falls. This is known as the Solvay Ammonia soda

process. The solution contains NH_4Cl, from which NH_3 may be recovered by treating it with CaO.

128. Dissolve 150 parts, by weight, of hyposulphite of soda in 15 parts boiling water, and gently pour it into a tall test-glass so as to half fill it, keeping the solution warm by placing the glass in hot water. Dissolve 100 parts, by weight, of sodium acetate in 15 parts hot water, and carefully pour it in the same glass; the latter will form an overlying layer on the surface of the former, and will not mix with it. When cool, there will be two supersaturated solutions. If a crystal of sodium hyposulphite be attached to a thread, and carefully passed into the glass, it will traverse the acetate solution without disturbing it, but, on reaching the hyposulphite solution, will cause the latter to crystallize instantaneously in large prisms. (Compare note, p. 134.)

129. Dissolve 40 grs. of common soda in one wine-glass of water, and 35 grs. of tartaric acid in another. On being poured together in a goblet, they will violently effervesce. Use a glass which is large enough to prevent any of the liquid from running over. Neatness in experiments is essential to perfection, and often to success. At the close of this illustration, evaporate the solution,* and a neutral salt will result. (Compare p. 98).

138. Calcium.—130. Place a few lumps of marble in the open fire, or in an open crucible with a hole at the bottom, and heat it strongly for an hour or two. It is converted into quicklime. Place the lumps in an open dish, and cover them with water—about one third the weight of the lime. The water will be all absorbed by the lime, and slaked lime $Ca(OH)_2$ produced.

139.—131. Put some of the $Ca(OH)_2$ in a bottle of water, and cork. The water dissolves a very little of the $Ca(OH)_2$, and the clear solution forms "lime-water." (Compare Ex. 54.) It may be decanted or siphoned off, and a fresh supply made by simply filling up the bottle again. This may be repeated as long as solid $Ca(OH)_2$ remains in the bottle.

132. Pour some of the lime-water into a saucer, and let it stand. It becomes covered with a film which may be shown to be a carbonate by removing it, placing it in a test-tube, and adding a drop of HCl. Effervescence will occur, and the gas which comes off will render turbid a drop of clear lime-water introduced into the mouth of the test-tube on the end of a glass rod.

141.—133. Select a medal suitable for the purpose; paste a shallow rim of paper round it, so as to make it like the lid of a pill-box, and anoint the surface of the medal very lightly with oil. Mix a little dry plaster of Paris

---

* Pour a part of the liquid into an evaporating dish, and place this on the tripod over the flame of the spirit-lamp, or upon a hot stove. Heat until a drop of the liquid, taken out on the end of a glass-rod and put on a bit of glass, will crystallize as soon as it cools. Then set the dish aside to cool, when crystals will soon begin to form.—In this connection, it is well to remark that a *cook stove will be found of great use in chemical experiments, and indeed may, in the laboratory. take the place of the furnace.* The oven will dry apparatus and chemicals; the heat is sufficient for evaporating solutions, distilling water, etc., while an excellent sand or water-bath may be readily contrived.

with water till it becomes of the consistence of thin cream; apply it carefully with a hair-pencil to every part of the surface, so as to exclude air-bubbles; then pour a thicker mixture into the mold. Allow it to remain for an hour. The cast may then be removed: it will be a reversed copy of the medal.

143. See *Experiment* 80.

134. Dissolve a little piece of marble in HCl (see Ex. 51). Clean one end of a thin platinum wire by dipping it in HCl, and then holding it in the flame of a Bunsen burner repeatedly until it imparts no color to the flame when first introduced. Then dip the wire into the $CaCl_2$ solution, and hold it in the flame. An orange-red flame coloration will be obtained, which is characteristic of calcium.

135. Repeat the last experiment with $SrCl_2$ and with $BaCl_2$.

144. *Magnesium.*—136. Burn a piece of magnesium ribbon. It is readily kindled by a candle or match.

148.—137. Repeat Ex. 134 with solutions of K, Na, Li, Cs, Cu. The wire must be carefully cleaned before making the experiment with each.

151. *Iron.*—138. Pulverize a salt of iron, and heat it with Na_2CO_3 on charcoal before the blow-pipe. A black powder is obtained, which is attracted by the magnet.

139. Mix some iron filings with twice the bulk of flowers of sulphur. Heat the mixture in a hard glass tube. The two will unite and form ferrous sulphide FeS, which yields H_2S, when treated with dilute H_2SO_4 (see p. 109).

157.—140. Make a strong solution of potassium permanganate, and heat to boiling in a test-tube. Pour in a few drops of glycerin; the latter is oxidized so violently that a flash may be seen, and part of the liquid is ejected from the test-tube.

161. *Copper.*—141. Fill a test-tube nearly full of H_2O. Pour in it a few drops of the solution of copper sulphate. Add NH_4OH, and a greenish-blue precipitate will be formed, which dissolves when more NH_4OH is added, yielding a dark-blue solution. The copper sulphate may be readily prepared for this experiment by heating a copper cent with strong sulphuric acid. This experiment may be made to show the divisibility of matter by weighing the cent, finding what proportion of the whole solution you use, and then experiment to see what quantity of water can be taken, and yet have the blue color perceptible in the ammonia test.

142. Besides the ammonia test for copper, the metal may be detected by the red metallic deposit formed on an iron nail, dipped into a solution of the salt.

143. Dissolve a piece of Cu in HNO_3 (see p. 33). Evaporate to a small quantity the solution which is obtained, and allow it to cool. If sufficiently concentrated, crystals of $Cu2NO_3$ will be formed.

144. Repeat Ex. 143 with Cu and strong H_2SO_4 (see p. 104). The blue crystals are $CuSO_4$.

163. *Lead.*—145. If a water contain lead, even in minute quantity, its presence is easily ascertained by taking two similar jars of 25 cm. high, of colorless glass, filling both of them with the water, and adding to one of the jars 3 or 4 cc. of a solution of sulphuretted hydrogen. A quantity of lead less than one part in two millions is easily perceived by the brown tinge occasioned, on looking down upon a sheet of white paper; the jar to which the test has not been added serving as a standard of comparison.

165. *Gold.*—146. Place a little gold-leaf in two test-tubes; to one add HNO_3, to the other HCl. Even when heated, the gold-leaf will remain unaffected in each. Pour the contents of one tube into the other: the Au will disappear with effervescence. Evaporate this solution in a small porcelain dish till the acid is nearly all driven off: gold chloride will be left.

147. Dilute the solution with 3 or 4 cc. of H_2O. To a portion of this liquid add a solution of ferrous sulphate: a brown precipitate of finely-divided reduced Au is obtained, and iron chloride is formed.

167. *Silver.*—148. Dissolve a ten-cent piece in HNO_3. The solution has a bluish color, owing to the presence of the Cu. Dilute with 200 cc. of water; then add a solution of NaCl so long as it forms a precipitate; white flakes of silver chloride are formed. Stir the mixture briskly with a glass rod; the precipitate of AgCl will collect into clots. Filter the solution. The presence of Cu in the clear liquor may be proved by adding to a portion of the liquid NH_4OH in excess: a blue solution is formed. Place the blade of a knife in another portion of the filtrate: it will become coated with metallic Cu.

149. Take the precipitated silver chloride, and, after having washed it well on a filter, place it in a wine-glass with a little water; add two or three drops of H_2SO_4, and then place a slip of Zn in contact with the chloride, and leave it for twenty-four hours. The chloride will be reduced to metallic Ag, which will have a gray porous aspect, while zinc chloride will be found in solution. Lift out the piece of Zn carefully; wash the Ag first with water containing a little H_2SO_4, then with pure H_2O. Dry the residue. Place a small quantity of it upon an anvil, and strike it a blow with a hammer; a bright metallic surface will be produced. Place a little of the gray powder upon charcoal, and heat in the flame of a blow-pipe: it will melt into a brilliant malleable bead. Dissolve another portion in HNO_3; red fumes will escape, and silver nitrate be obtained in solution.

150. Fill a vial half full of a solution of silver nitrate, and add a few globules of Hg. The Ag will be precipitated in a few days, forming the "silver tree."

151. Place a crystal of $AgNO_3$ on a piece of charcoal, and heat in the reducing blow-pipe flame. A globule of metallic Ag is produced, which is soft and malleable.

152. Float a sheet of sized paper on an NaCl solution. When dry, float it for three minutes on a solution of $AgNO_3$, and dry in the dark. Press a fern-leaf on a piece of glass, lay a sheet of this paper on it, and then a thin board as large as the glass. Clamp them together with clothes-pins, and expose to direct sunlight. When sufficiently black, place in a solution of "hypo," and wash thoroughly with water.

ORGANIC CHEMISTRY.

195.—153. Dissolve about 20 grams of grape-sugar in a liter of water, and put the solution, with a little yeast, in a flask connected with a cylinder or bottle, as shown in Fig. 52, *A* and *B*. The cylinder *B* contains some clear lime-water, and the end of the short, bent tube is plugged with cotton to prevent the air from entering freely. After standing some hours in a warm room, *B* will contain a white precipitate of $CaCO_3$, formed by the CO_2 produced by the fermentation. The contents of *A* will have an alcoholic odor, and, by distilling, a dilute alcohol could be obtained.

200.—154. Pour a little alcohol into a small beaker, and suspend over it a coil of Pt wire, which has been heated until it glows. It will continue to glow, owing to a slow oxidation of the alcohol vapors. The peculiar penetrating odor observed is due to *aldehyde*.

155. Another instructive experiment by which aldehyde is produced is as follows: Add to a solution of potassium bichromate, sulphuric acid and a little alcohol. On heating, the odor of aldehyde will be perceived, while the red solution becomes green. Explain.

156. Formic acid may be made by distilling a mixture of one part of oxalic acid and ten parts of glycerin. The dilute acid thus obtained may be used for making copper formate: Dissolve in the acid freshly precipitated and washed copper oxide until no more is taken up; filter and concentrate the solution, if necessary, by evaporation, when copper formate will separate in beautiful crystals.

203.—157. Make oxalic acid by pouring ordinary strong nitric acid upon sugar (6 pts. of acid to 1 of sugar), and heating gently till the reaction begins. The sugar is oxidized with some violence and evolution of abundant red fumes. On cooling, oxalic acid will crystallize out. More can be obtained by evaporation. Purify the acid by recrystallizing it from water.

205.—158. Ether may be made by heating to 140° C. a mixture of 100 grams of alcohol and 180 grams concentrated sulphuric acid. The operation should be conducted in a retort connected with a cooler.

206.—159. Mix in a test-tube equal parts of acetic acid and alcohol, and add a little concentrated H_2SO_4. The odor is that of *ethyl acetate*, or acetic ether.

210.—160. Add to a solution of sodium carbonate a little alcohol, and then, after warming it to about 80° C., introduce gradually a few crystals of iodine. Yellow crystals of iodoform will separate from the solution.

219.—161. To one gill of water, add 15 or 20 grains of strong H_2SO_4. Place in a large flask, and heat. While boiling, drop in slowly two drams of starch, finely powdered. Boil for several hours, adding water as may be necessary. Finally, drop in slowly fine chalk until the liquid is neutral; then cool, filter off the calcium sulphate. The solution contains grape-sugar. Test it as described in the next two experiments.

162. Put a little AgNO$_3$ solution in a test-tube, and add aqua ammonia slowly until the brown precipitate has again dissolved. Pour some of this into a second test-tube, and add a solution of grape-sugar. On boiling, the silver will be reduced in the form of a brilliant mirror on the side of the tube.

163. Take a solution of CuSO$_4$, and add enough tartaric acid to prevent its being precipitated by KOH. Then add enough KOH solution to make it strongly alkaline. Boil this solution, and add, drop by drop, a dilute solution of grape-sugar. The intensely blue color disappears, and a red precipitate of Cu$_2$O is produced. This is called *Fehling's test*, and is employed to detect sugar in diabetic urine. By taking proper precautions, the amount of the sugar may be accurately determined.

222.—164. Mix together in a test-tube 2 parts of H$_2$SO$_4$ and 1 part of HNO$_3$, and, when cold, allow some benzene to flow into it, a few drops at a time, waiting each time till the reaction is complete, and keeping the mixture cool. On pouring the mixture into water, it will be found that the benzene no longer floats on water as before, and has acquired a very agreeable odor. An atom of H has been replaced by NO$_2$, forming *nitro-benzene*, C$_6$H$_5$NO$_2$. The vapor of the nitro-benzene should not be inhaled.

165. Dissolve a little aniline in water, and add a filtered solution of bleaching powder. A purple color is produced.

237.—166. Make a dilute solution of gelatin, and add a solution of tannic acid; a leathery precipitate is formed.

240.—167. Fill a test-tube one third full of fresh milk, and add an equal volume of water, then a little acetic acid (or rennet), and allow it to stand a short time; filter out the precipitate, and wash with water. The precipitate consists of casein and fats. The filtrate contains sugar, as may be shown by its reducing action on an alkaline solution of tartrate of copper. Remove the precipitate from the filter, and shake it up in a test-tube with ether. This dissolves out the fat. Filter again; let the filtrate evaporate, and the fat will be left in a pure state.

168. Evaporate a larger portion of milk to dryness, and heat until the residue is quite white. Dissolve in water, filter, and test for chlorides with AgNO$_3$.

243.—169. Mix intimately 10 grains of gelatin with 50 grains of soda-lime, and heat strongly; NH$_3$ gas will be evolved, and can be detected by its odor, its action on reddened litmus-paper, and by fuming with HCl.

250.—170. The ordinary photographic process, as given on p. 171, is a good illustration of the power of the sun's rays.

171. Dissolve 1 gram of ammonia-citrate of iron in 20 cc. of water; add to it 20 cc. of a solution of K$_6$FeCy$_6$, made in the same manner. Keep the mixed solution in the dark. Float a sheet of white paper on the solution, and allow it to dry. Cover a plate of glass with black varnish, and before it dries write upon it with any sharp-pointed instrument. When perfectly dry, place it over a sheet of the prepared paper, and expose two or three hours to bright sunlight. Remove, and wash well in cold water. You will have the writing in blue on a white ground.

172. Mix together equal volumes of a solution of gelatin and a solution of $K_2Cr_2O_7$, and add a little tincture of logwood. Pour it into a flat dish, and float upon it a sheet of unruled writing-paper, and dry in the dark. Expose under a negative (p. 172), or fern leaves, to direct sunlight for an hour, then wash thoroughly with *warm* water. Bichromated gelatin is rendered insoluble by exposure to sunlight. This principle is employed in all photo-engraving processes.

173. Mix together equal volumes of H and Cl in the dark, fill a small glass bulb, and throw it up into a bright beam of the sun, when it explodes *with violence*. Do not at any time hold the bulb in the hand, or place it within several feet of the eyes.

174. Plants decompose CO_2 in the sunlight. (See p. 89 and Exp. 70.)

QUALITATIVE ANALYSIS,

FOR BEGINNERS.

[The following pages on analysis were prepared by Prof. E. J. HALLOCK, Ph.D., of Boston.]

IN order to be able to analyze almost every inorganic substance met with in the arts, or sold in the shops, it is only necessary for the student to familiarize himself with the reactions of about twenty-six metals and a dozen acids. To be able to apply these tests with certainty, in all cases, and to know the easiest and best methods of dissolving the substance, constitute a qualitative chemist.

For reasons which will appear farther on, metals are divided into five groups.

THE FIRST GROUP embraces lead, silver, and the mercurous salts of mercury. They are classed together because they are the only metals which are precipitated from solution by hydrochloric acid. The student should take a solution of lead nitrate, Pb $2NO_3$, formed by dissolving litharge or lead in nitric acid, or some lead acetate solution (see page 164), and try the following tests, making a note of his results. With HCl a white precipitate of $PbCl_2$ is formed. This precipitate is filtered out and washed. It dissolves in boiling water, and crystallizes from this solution on cooling. To another portion of the solution add H_2SO_4; a white precipitate of $PbSO_4$ is produced, which is insoluble in H_2O. To a third portion add potassium chromate, K_2CrO_4; a yellow precipitate is formed. To another portion add KI; a yellow precipitate is again produced, but it dissolves in boiling water, and forms beautiful crystals on cooling.

Repeat each of these tests with silver nitrate, $AgNO_3$ (*experiment* 149). The precipitate with HCl is insoluble in boiling H_2O, but dissolves in NH_4OH, and is reprecipitated by HNO_3. With KI a yellowish-white precipitate is formed.

Repeat with mercurous nitrate, $HgNO_3$, made by the action of dilute HNO_3 on an excess of Hg, in the *cold*. We have with HCl a precipitate of calomel ($HgCl$), (page 176), which is insoluble in H_2O, and blackens on adding NH_4OH, but does not dissolve. KI forms a greenish-yellow precipitate.

SEPARATING METALS OF GROUP I.—Mix the solutions of the three metals, and add HCl. Filter, and boil the precipitate in water; filter hot, and to the *filtrate* add K_2CrO_4. The yellow precipitate proves that lead is present. Boil the residue in ammonia and filter; to the filtrate add HNO_3. A white precipitate proves silver present. The black insoluble residue is a compound of mercury (Hg_2H_2NCl).

THE SECOND GROUP embraces the mercuric salts of mercury, together with Pb, Bi, Cu, Cd, As, Sb, Sn, Au, and Pt. They are precipitated from acid solutions as sulphides by passing H_2S gas through the solution. (See *experiment* 103.) Of these HgS, PbS, Bi_2S_3, CuS, and CdS are insoluble in ammonium sulphide, $(NH_4)_2S$, and constitute the first division of this group. The sulphides of the remaining five metals are soluble in $(NH_4)_2S$, * and form the second division.

Pass H_2S gas into a solution of corrosive sublimate ($HgCl_2$) (page 176); a precipitate is formed which is at first white, then yellow, red, and finally black. It is insoluble in $(NH_4)_2S$, and in HNO_3. It dissolves in aqua regia and gives a gray precipitate with an excess of $SnCl_2$. Repeat the first experiment with some lead solution; a black precipitate is formed, soluble in boiling HNO_3. In this solution a white precipitate of $PbSO_4$ is formed on adding H_2SO_4. Add a few drops of KI solution to the original solutions of $HgCl_2$, and $Pb2NO_3$; in the former case the precipitate (HgI_2) is red, in the latter (PbI_2) it is yellow. These tests are characteristic of the metals when alone. Pass H_2S

---

* $(NH_4)_2S$ is prepared, according to Fresenius, by saturating a given volume of ammonia solution (specific gravity 0.96) with H_2S gas, and adding to it an equal volume of the ammonia. The solution, which is at first colorless, soon becomes yellow by keeping, or may be at once converted into the yellow sulphide by the addition of sulphur. It should yield no precipitate with magnesium sulphate (Epsom salt). This re-agent is decomposed by acids, sulphur being precipitated.

into Bi3NO$_3$ solution;* a black precipitate is formed; dissolve in HNO$_3$; add a drop of H$_2$SO$_4$ to prove it is not lead; then cautiously add ammonia, which produces a white precipitate of Bi(OH)$_3$. Repeat all the above experiments with CdSO$_4$ solution; the precipitate with H$_2$S is a beautiful yellow, soluble in HNO$_3$, but insoluble in KCy. Pass H$_2$S in CuSO$_4$ solution, and a brownish-black precipitate will be formed, soluble in HNO$_3$ and in KCy. Salts of copper have a bluish color, which becomes more intense on adding NH$_4$HO. (*Experiment* 141.) With potassium ferrocyanide (K$_4$FeCy$_6$) they give a reddish-brown precipitate insoluble in HCl.

SEPARATING METALS OF SECOND GROUP, FIRST DIVISION.—After the student has made all the above reactions he may mix the solutions of the five metals and proceed to separate them. Some of the lead is precipitated by HCl, and is filtered out before H$_2$S is passed through the solution. The precipitate with H$_2$S is boiled in HNO$_3$, and HgS remains as a residue. When the original solution was very acid, S will be found mixed with HgS in the residue insoluble in HNO$_3$. To the filtrate add a little H$_2$SO$_4$ to precipitate the lead present; filter from the PbSO$_4$ and add NH$_4$OH; Bi(OH)$_3$ is precipitated. The precipitate, dissolved in aqua regia and concentrated by evaporation, should give a white precipitate if poured into water. The addition of NH$_4$Cl aids this reaction. The *blue* filtrate from the Bi precipitate is boiled with KCy, care being taken not to inhale the fumes, and H$_2$S added; CdS forms a yellow precipitate. The presence of Cu in the filtrate from CdS is proved by the formation of a reddish-brown precipitate with K$_4$FeCy$_6$, after HNO$_3$ has been added to the solution.

The most interesting metal of *group second, second division,* is As. H$_2$S in an HCl solution of As$_2$O$_3$ gives a yellow precipitate As$_2$S$_3$; it is soluble in (NH$_4$)$_2$S, and in (NH$_4$)$_2$CO$_3$. The neutral salts of arsenious acid yield with AgNO$_3$, a yellow precipitate,

* When Bi solutions are diluted with water, basic salts are precipitated unless there be much free acid present. This reaction is most sensitive with BiCl$_3$, so that HCl may be used to dissolve the Bi3NO$_3$ for the H$_2$O test.

Ag_3AsO_3, soluble in HNO_3. A small piece of bright green paper often contains enough of this metal to give several characteristic tests. Apply a single drop of nitric acid to the paper; a moment after neutralize with ammonia and observe the color, a deep blue always indicating copper. When the white fumes have nearly disappeared, apply to the same spot a drop of $AgNO_3$; a yellow ring indicates As. The most delicate test for As as well as Sb is Marsh's test (see page 119). The mirror formed by As on porcelain is soluble in sodium hypochlorite (Labarraque's solution); that of Sb is insoluble in this. If AsH_3 is passed into $AgNO_3$, metallic Ag is precipitated and enough HNO_3 is set free to keep the As in solution until it is carefully neutralized with NH_4OH, when a precipitate of Ag_3AsO_3 appears.

Antimony closely resembles As in its reactions. Pass H_2S into a solution of tartar emetic (*experiment* 103), and an orange-colored precipitate, Sb_2S_3, will be formed, soluble in $(NH_4)_2S$, and re-precipitated on adding dilute HCl to this solution. This precipitate is soluble in strong HCl, while the corresponding arsenic precipitate is not. Put some of the Sb solution in a new Marsh's apparatus. The mirror is insoluble in sodium hypochlorite.

Tin dissolves in HCl, but is oxidized to a white powder by HNO_3 without dissolving. There are two series of tin salts; $SnCl_2$ gives a dark brown precipitate with H_2S; $SnCl_4$ a yellow precipitate with H_2S, both soluble in yellow $(NH_4)_2S$. $SnCl_2$ forms with an excess of $HgCl_2$, a white precipitate of $HgCl$; but when $SnCl_2$ is in excess a gray precipitate of Hg is formed. On adding dilute HCl to the $(NH_4)_2S$ solution of the sulphide, it is re-precipitated yellow, even though it may have been brown before it was dissolved. It dissolves in strong HCl.

Gold and platinum are distinguished from all other metals by their insolubility in HCl or HNO_3, but are converted into soluble chlorides by aqua regia. The characteristic test for Au salts is $SnCl_2$ mixed with $SnCl_4$, the purple of Cassius being formed. $PtCl_4$ will give a yellow precipitate with NH_4Cl and alcohol, $(NH_4)_2PtCl_6$.

SEPARATING METALS OF SECOND GROUP, SECOND DIVISION.—Into an acid solution of Sn, Sb, and As, pass H_2S gas. Filter and dissolve precipitate in $(NH_4)_2S$ to remove any members of first division, if these are to be looked for. Re-precipitate the sulphides with dilute HCl, filter and wash. Then treat with strong hot HCl ; if a residue remains it is probably As_2S_3. Dissolve this in HCl with the aid of a little solid $KClO_3$, and test, as described in *experiment* 110, or in Marsh's apparatus. The filtrate from As_2S_3 contains Sb and Sn. Put this in another Marsh's apparatus ; the mirror will be insoluble in hypochlorites. When the zinc has all dissolved, take the residue and dissolve it in HCl and test for Sn with $HgCl_2$.

As Au and Pt are seldom to be sought for, this part of the separation may be omitted. The mixed sulphides of Au, Pt, As, Sb, and Sn may all be placed in Marsh's apparatus, as directed in the table on page 302, when the As and Sb combine with H, and are separated by passing the AsH_3 and SbH_3 into $AgNO_3$. The metallic Sn, Au, and Pt remain in the H generator ; the Sn is then dissolved out with HCl, and tested with $HgCl_2$; the Au and Pt are dissolved in aqua regia and tested in separate portions of the solution, as described above.

GROUP THIRD embraces Co, Ni, Fe, Cr, Mn, Al, and Zn. They are precipitated by $(NH_4)_2S$ from neutral or alkaline solutions. The characteristic test for Co, is the blue color imparted to a borax bead. Ni alone gives, in the outer blow-pipe flame, a reddish-brown bead, in the inner gray. Both give with $(NH_4)_2S$ black precipitates insoluble in dilute HCl. If KNO_2 and acetic acid are added to a solution of Co and Ni, the former is slowly precipitated and not the latter. To a solution of $FeSO_4$, add a drop of potassium ferricyanide (K_3FeCy_6) ; a blue precipitate is formed. To another portion add $(NH_4)_2S$; a black precipitate is formed, soluble in dilute HCl, from which solution it is re-precipitated by NaOH, as greenish $Fe(OH)_2$. Repeat the latter test with ferric chloride (Fe_2Cl_6) and reddish brown $Fe_2(OH)_6$ is precipitated. Fe_2Cl_6 gives with K_4FeCy_6 a precipitate of Prussian blue. In a glass of water place one drop of Fe_2Cl_6, and add a few drops of potassium sulphocyanide (KCyS) ; the liquid acquires a blood-red color. Iron salts also give characteristic colors to the borax beads ; yellow in the outer and green in the inner flame. To a solution of $MnSO_4$, add $(NH_4)_2S$; a flesh-colored precipitate is formed, soluble in HCl ;

NaOH precipitates $Mn(OH)_2$. The borax bead with Mn acquires an amethyst-red color (see note, page 122) in the outer blow-pipe flame. Fused with Na_2CO_3 and KNO_3 a green mass is formed. $(NH_4)_2S$ gives with the salts of Cr, a greenish precipitate of $Cr_2(OH)_6$, soluble in HCl, and re-precipitated when boiled with NaOH. Fused with Na_2CO_3 and KNO_3 it forms yellow potassium chromate K_2CrO_4. When this is dissolved in water and acidified with acetic acid, it yields a yellow precipitate with Pb2NO₃. To a solution of alum add $(NH_4)_2S$; it forms a white precipitate soluble in dilute HCl. To the second portion add NH_4OH, a white precipitate; and to a third add NaOH and boil; the white precipitate at first formed dissolves again. To a solution of $ZnSO_4$, obtained in *experiment* 30, add NH_4OH slowly. The white precipitate at first formed dissolves when more NH_4OH is added. In this solution H_2S causes a white precipitate of ZnS.

SEPARATION OF METALS OF GROUP THIRD.—Some NH_4Cl and NH_4OH is first added; then $(NH_4)_2S$. The precipitate is digested in dilute HCl; Ni and Co are sought in the residue. The filtrate is boiled with NaOH for half an hour in a porcelain dish, and filtered. To the filtrate HCl is added until acid, then NH_4OH, which precipitates Al; Zn is precipitated from the filtrate from this by H_2S or $(NH_4)_2S$. The residue, containing Fe, Mn, and Cr, is fused with pure KNO_3 and Na_2CO_3; if the mass is green, Mn is indicated; if yellow, Cr. One half of the mass is warmed with water; the insoluble residue is tested for iron; the filtrate is tested for Cr by first neutralizing with acetic acid, and then adding Pb2NO₃. The test for Mn is to boil some of the fused mass in HNO_3, with red lead; a beautiful rose pink is produced from the formation of $KMnO_4$.

GROUP FOURTH embraces the metals of the alkaline earths, Ba, Sr, and Ca, whose carbonates, precipitated by $(NH_4)_2CO_3$, are insoluble in H_2O, but soluble in acetic acid or in HCl. $BaCl_2$ forms with H_2SO_4 or a clear solution of $CaSO_4$, a precipitate insoluble in acids; K_2CrO_4 precipitates yellow $BaCrO_4$. Ba compounds impart a green color to the flame of an alcohol lamp or Bunsen burner. $CaCl_2$, with ammonium oxalate, yields a white precipitate insoluble in acetic acid. H_2SO_4 produces a white precipitate of $CaSO_4$, slightly soluble in water and acids. Ca

salts color the flame yellowish red. $SrCl_2$ gives a white precipitate with a clear solution of $CaSO_4$; if the solution is dilute,
half an hour is required for the precipitation. Sr colors the
flame crimson-red.

SEPARATING METALS OF GROUP FOURTH.—Some NH_4Cl is first
added, if not already present in the solution, then $(NH_4)_2CO_3$.
The precipitate is dissolved in acetic acid, and divided in two
portions. To one is added K_2CrO_4; Ba is precipitated yellow,
and the filtrate is tested for Sr by means of $CaSO_4$. H_2SO_4 is
added to the other portion, the precipitate filtered out. NH_4OH
is added to the filtrate, which is now tested for Ca with ammonium oxalate.

GROUP FIFTH embraces Mg, K, Na, and Li. As lithia is very
rare, we omit its reactions. $MgSO_4$ yields a white precipitate of
$Mg(OH)_2$ on the addition of NH_4OH, unless the solution contains
NH_4Cl; hence the necessity of adding NH_4Cl, before testing for
groups III. and IV., where NH_4OH would otherwise throw down
Mg. The salts of Mg give à white precipitate with sodium
phosphate, Na_2HPO_4, and NH_4OH. KCl yields a yellow precipitate with $PtCl_4$; Potassium tartrate is precipitated from concentrated solutions by sodium tartrate. K_2SO_4 gives a white
precipitate with $2HF,SiF_4$ and alcohol. K imparts a violet color
to flame, which appears reddish violet when viewed through
blue glass. Na is not precipitated by any of the above re-
agents; it imparts an intense yellow color to flame. K, Na, and
Li, as well as Ca, Ba, and Sr, are easily detected by the spectroscope.

Ammonia is liberated from its compounds by heating with
NaOH or $Ca(OH)_2$, and is then recognized by the smell, by bluing
red litmus, and by producing white fumes when a rod moistened
with HCl gas is held over it. (See page 35.)

TESTS FOR ACIDS.

The acids do not admit of the strict grouping and successive
separation employed for metals, and we will rest content with
mentioning the simplest tests for the principal acids, beginning
with those of the halogens:

HCl with $AgNO_3$, white precipitate, soluble in NH_4OH, not in HNO_3.

HI with $AgNO_3$, yellowish precipitate, almost insoluble in NH_4OH.

HI with $HgCl_2$, red precipitate, soluble in KI.

HI with starch paste and Cl solution or bleaching powder, blue color. (See page 101.)

CaF_2 with H_2SO_4 liberates HF, which attacks glass. (See page 102.)

HBr with starch paste and Cl water, an orange-yellow color.

H_2SO_4 with $BaCl_2$, white precipitate, insoluble in HCl.

SiO_2 is insoluble in H_2O, as are most of the silicates except those of K and Na. In analyzing the soluble silicates, they are first evaporated to dryness with excess of HCl, the soluble chlorides dissolved in H_2O or HCl, and the SiO_2 left as a gritty powder.

Boric Acid is detected by placing it in a dish containing alcohol and H_2SO_4, and igniting the alcohol. A green tinge to the flame indicates B. If a mixture of glycerin and borax is brought into a flame and removed as soon as it takes fire, the green flame is easily recognized. (*Experiments* 112 and 113.)

H_3PO_4 with neutral $AgNO_3$, yellow precipitate, soluble in HNO_3 and NH_4OH.

H_3PO_4 with solution of ammonium molybdate in HNO_3, a fine yellow precipitate.

H_3PO_4 with $MgSO_4$ solution containing NH_4Cl and NH_4OH, a white precipitate soluble in acids. (See test for Mg, page 294.)

CO_2. Carbonates effervesce on the addition of acids, CO_2 being set free, which extinguishes a match inserted in the test-tube. The ear is often able to detect slight effervescence not otherwise perceptible.

HNO_3 is not precipitated by any re-agent. Into a test-tube containing some nitrate, drop a crystal of $FeSO_4$, then allow a few drops of H_2SO_4 to flow down the side of the test-tube which is held inclined. A characteristic dark-brown ring forms immediately. If Cu and strong H_2SO_4 are heated with a nitrate, red fumes are given off. A nitrate heated on charcoal deflagrates.

Chlorates deflagrate more violently than nitrates. H_2SO_4 liberates Cl_2O_4, which is betrayed by its color and odor. If a crystal of $KClO_3$ and a piece of P be placed in a glass of water, and a drop of H_2SO_4 conveyed to it by a pipette or tube, the P takes fire and burns under water (page 131). All experiments with chlorates must be performed with minute quantities, because of the great danger of explosions.

SO_2 is easily recognized by its odor. When sulphites are treated with HCl, the SO_2 is evolved.

If Cl gas be given off on heating a substance in HCl, the presence of a binoxide may be suspected. (See page 92.)

If an odor of H_2S is perceived on treating a substance with HCl, it is evidently a sulphide. Heated with strong HNO_3, sulphides are converted into sulphates.

HCy with $AgNO_3$, white precipitate soluble in KCy; difficultly soluble in NH_4OH. Care must be taken in handling the poisonous cyanide. On adding HCl to a cyanide, HCy is liberated, and is detected by the odor, which resembles bitter almonds.

H_4FeCy_6 with $AgNO_3$, white precipitate insoluble in NH_4OH, and in HNO_3. With Fe_2Cl_6, Prussian blue $[Fe_4(FeCy_6)_3]$ is formed.

Oxalic Acid, $H_2C_2O_4$, yields a white precipitate with $CaCl_2$, which is insoluble in acetic acid.

When several acids are present they are tested for in separate portions. As_2O_3 may occasionally be mistaken for H_3PO_4. $BaCl_2$ will yield precipitates with carbonic, oxalic, phosphoric, and sulphuric acids, but they all dissolve in HCl except $BaSO_4$.

To test for HCl in the presence of HI, or HBr, or both, the dry powder is mixed with dry $K_2Cr_2O_7$, and pure concentrated H_2SO_4 added, and heated, when CrO_2Cl_2 is given off as a brownish-red gas; and a glass rod dipped in NH_4OH and held over the tube becomes slightly yellow if Cl is present, from the formation of $(NH_4)_2CrO_4$. If the substance to be tested contains only Br and I, these two may be separated by $CuSO_4$ and H_2SO_3, which precipitates the latter as Cu_2I_2 and permits the use of the starch test for Br. Or chloroform is added to the solution and a very little Cl water, or solution of bleaching powder, is then poured in and the test-tube well shaken. Then I is liberated and dissolved by the chloroform, imparting to it the well-known purple color, the intensity of which conceals the yellowish-brown color of the Br likewise set free and dissolved. On adding more Cl water, the violet disappears, leaving only the yellow color due to the Br.

PRELIMINARY TESTS.

A few tests in a dry way will give some clew to the substances present; but in a complete analysis *every* acid and *every* metal must be sought for.

I. HEATING IN A TUBE OF HARD GLASS CLOSED AT ONE END.— If the substance blackens, organic matter is probably present. If vapors escape, they are tested for CO_2, SO_2, H_2S, etc. If a sublimate is formed, it may be S, I, Hg, a compound of Hg, As, Sb, or a salt of ammonium.

II. HEATING ON CHARCOAL.—A small portion of the substance is placed on charcoal and exposed to the inner blow-pipe flame. (See page 83.) If an infusible white residue remains, moisten it with Co2NO₃ and heat it again; a fine blue indicates Al, a phosphate or SiO_2, a reddish tint Mg, a green color Zn. Mix another portion with Na_2CO_3 and heat on charcoal in the reducing flame. If a metallic globule is formed without an incrustation, it indicates Au, Cu, or Ag, as the color is yellow, red, or white. A very fusible and malleable globule surrounded by a yellow incrustation indicates Pb ; if the incrustation is white it may be Sb or Sn ; if orange, yellow while hot, becoming lighter on cooling, Bi. The globules of Sb and Bi are hard and brittle. If As is present, an odor resembling garlic is noticed. The charcoal tests will be of little use to the student, except for detecting Ag and Pb, until practice has given him considerable facility in the use of the blow-pipe.

III. BORAX BEADS.—Several metals impart characteristic colors to borax glass when fused with it before the blow-pipe. The end of a piece of platinum wire is bent to form an eye as large as this letter O ; it is next dipped in borax and held in the flame until fused, then dipped in the powdered substance and fused again. The color varies according as the oxidizing or reducing flame is employed.

Color.	Oxidizing.	Reducing.
Blue	Co and Cu	Co
Green	Cr and Cu	Fe and Cr
Red	Fe and Ni	Cu (opaque)
Amethyst	Mn	—
Yellow to brown	Fe	—
Colorless	Si, etc.	Mn
Gray	—	Ni

SOLUTION.

The first thing to be done before beginning an analysis is to bring the substance into solution. Distilled water is first employed; if a residue insoluble in water remains, it is treated with acid. In analyzing metals and alloys, nitric acid is the usual solvent; aqua regia being required only for the noble metals. If Sn is present, and Pb and Ag absent, HCl is employed. Mineral substances, if insoluble in any acid, are rendered soluble by *fluxing*, or fusing with pure Na_2CO_3 and K_2CO_3. As a very high heat is required for fluxing, *deflagration* is sometimes preferred. One part of the insoluble powder is intimately mixed with two parts of dry sodium carbonate, two parts pulverized charcoal, and twelve parts niter. The mixture is placed in the open air and a match applied. A portion of the porous mass produced will be soluble in water, the remainder in acids. The two solutions are to be preserved and tested separately. The metals will be found in the acid solution, while the acids will be found in the aqueous solution. Before beginning the regular course of analysis with these solutions, part of the aqueous solution is evaporated to dryness with excess of HCl to render all the SiO_2 insoluble. In separate portions of the aqueous solution, the various acids are sought as above described (p. 294).

If a portion of the substance is insoluble in HCl after fluxing, it is probably silicic acid, or an undecomposed silicate, and may be rendered soluble by fluxing a second time.

A platinum crucible must never be employed if reducible metals, especially Pb, As or Sb have been found in the preliminary tests.

EXAMPLES FOR PRACTICE.

After the student has made all the tests above given, and succeeded in separating the members of each group from each other, especial care being given to the separation of lead from bismuth, copper from cadmium, arsenic from antimony, and nickel from cobalt, the teacher may give out the following or similar substances for analysis, not following the precise order of the book, so that the student shall not know what substance he is analyzing. Each student should record the results of every analysis in a note-book which he will rule for each analysis as shown under No. 1.

1. ANALYSIS OF $CuSO_4$.—A crystal of this salt as large as a pea is given to a student, who dissolves it in distilled water in a test-tube and divides the solution in two portions. To one is added a drop of HCl, which should produce no precipitate. H_2S is then added until all the Cu is precipitated. The solution is then filtered and the precipitate thoroughly washed on the filter. H_2S should produce no precipitate in the filtrate. The precipitate, which is found to be insoluble in $(NH_4)_2S$, is dissolved in HNO_3, and no residue, except perhaps a little sulphur, remains, so that the absence of Hg is established. H_2SO_4 produces no precipitate in this solution, hence Pb is absent. Ammonia, added in excess, gives no precipitate (Bi is absent), but the intense blue color characteristic of Cu, and as only one metal is to be sought, the presence of Cu is further proved by adding HNO_3 and K_4FeCy_6, which causes a reddish-brown precipitate. The second portion of the solution is used in testing for acids. To a small quantity of this some $BaCl_2$ is added, and if the precipitate is insoluble in HCl, the acid present must be H_2SO_4. The results are recorded in tabular form thus:

ANALYSIS NO. 1.

SUBSTANCE BLUE, SOLUBLE IN H_2O.

GROUP I.	GROUP II.	GROUP III.	GROUP IV.	GROUP V.
HCl	H_2S	$(NH_4)_2S$	$(NH_4)_2CO_3$	
0	Brown precip., sol. in HNO_3. NH_4OH blue. K_4FeCy_6 brown. Cu.	0	0	0

ACIDS: $BaCl_2$; White precipitate insol. in HCl .... H_2SO_4.

2. ANALYSIS OF $HgCl_2$.—This salt is likewise very soluble in H_2O. To one portion of the solution add HCl, and H_2S. The latter produces a black precipitate, insoluble in HNO_3, which indicates Hg, but a confirmatory test must be employed, which is to dissolve the precipitate in aqua regia and add $SnCl_2$. To a second portion add some $BaCl_2$, which will cause no precipitate. To a third portion add $AgNO_3$, which produces a white precipitate, insoluble in HNO_3, but soluble in NH_4OH, proving the presence of HCl.

3. ANALYSIS OF $FeSO_4$.—Acidify a portion of the solution with HCl, add a little H_2S to prove that no metals of the second group are present, and then $(NH_4)_2S$, which produces a black precipitate, which is treated as directed for Group III., page 293. To some of the original solution a drop of K_3FeCy_6 is added, when the blue color proves the presence of Fe.

4. ANALYSIS OF $Sr2NO_3$.—Dissolve in water, test for Groups I., II., and III., which may occur as impurities, and then add $(NH_4)_2CO_3$. The white precipitate is filtered out and washed, then dissolved in acetic acid. To one portion add K_2CrO_4, when the absence of Ba is shown, and the Sr test may next be made, by adding NH_4OH and $CaSO_4$. The precipitate forms slowly. In the original solution no precipitate is formed by $BaCl_2$ or $AgNO_3$, and a careful test for HNO_3 is made with $FeSO_4$ and H_2SO_4, as described on page 295.

5. ANALYSIS OF $BaSO_4$.—This substance refuses to dissolve either in H_2O or in acids. It is boiled repeatedly with fresh quantities of Na_2CO_3 and filtered boiling hot; or fluxed with KNO_3 and Na_2CO_3. The filtrate contains Na_2SO_4; the residue is $BaCO_3$ (and unaltered $BaSO_4$). The residue is dissolved in HCl and tested for Ba with K_2CrO_4 or $SrSO_4$ solution.

6. ANALYSIS OF A COIN.—A silver coin is dissolved in HNO_3, then diluted and the Ag precipitated with HCl as $AgCl$. From this metallic silver is precipitated on a piece of clean zinc placed in the precipitate, which is moistened with dilute H_2SO_4. In the blue filtrate will be found all the copper, which may be tested for as above. If, instead of a silver coin, a nickel coin is used, HCl will give no precipitate, the Cu will be thrown down by H_2S, and the Ni by $(NH_4)_2S$. In analyzing compound substances, great care must be taken that all the metals of a certain group are precipi-

tated before proceeding to the next, and for this purpose, after precipitating the Ag with HCl, a drop of HCl is added to the filtrate to ascertain whether any Ag remains in solution. Precipitates should also be well washed, but the wash-water is not added to the filtrate.

7. ANALYSIS OF MIXED SALTS.—A mixture of Pb2NO$_3$, Bi3NO$_3$, Co2NO$_3$, KNO$_3$ may be dissolved in water and the metals sought in the above order (viz. Pb, Bi, Co, K). In testing for acids the student will remember that if Pb was found among the metals, H$_2$SO$_4$ and HCl must have been absent, as either would have precipitated the lead. If the student forgets this and adds BaCl$_2$ to the solution, it will form a precipitate of PbCl$_2$, which he might mistake for BaSO$_4$, and hence incorrectly suppose H$_2$SO$_4$ to be present.

Mixtures of various other soluble salts should now be given out, such as FeSO$_4$, NaCl, CuSO$_4$, and NH$_4$Cl; gradually increasing the number of metals and acids to be sought.

8. ANALYSIS OF LIME-STONE.—Dissolve any piece of marble or lime-stone in HCl. It will not be necessary to test for groups I. and II.; a small portion of the solution is tested with K$_4$FeCy$_6$ for iron. The alumina generally present is precipitated, along with the iron, by NH$_4$OH, after adding NH$_4$Cl, and is filtered out as rapidly as possible. In a small portion of the filtrate tests are made for Ba and Sr, which are of course absent, so that all the Ca may be precipitated by oxalate of ammonia. In the filtrate Mg will be found on adding Na$_2$HPO$_4$. The principal acid present is CO$_2$, as indicated by the effervescence with HCl when first dissolved. If a residue remained insoluble in HCl, it is probably SiO$_2$, or some silicate, and must be fluxed with Na$_2$CO$_3$ and K$_2$CO$_3$.

TABLE I.—SCHEME FOR

Add HCl *to Solution.*

PREC.   **FILTRATE.**

Ag Pb Hg Boil in H$_2$O	Add H$_2$S. **PRECIPITATE.**

Sol.	Prec.		
Pb	Ag Hg NH$_4$OH	Hg Pb Bi Cd Cu As Sb Sn Au Pt. Digest with yellow (NH$_4$)$_2$S.	

Residue.   *Solution.*

Sol.	Prec.		
With K$_2$CrO$_4$ yellow.	Ag Hg With HNO$_3$, white. Black.	Hg Pb Bi Cd Cu Boil in HNO$_3$.	As Sb Sn Au Pt In H apparatus with Zn & H$_2$SO$_4$.

Res.   *Sol.*   *Gas.*   *Residue.*

Hg.	Pb Bi Cd Cu With H$_2$SO$_4$.	As Sb Pass into Ag NO$_3$	Sn Au Pt Boil in HCl.

Prec.   *Sol.*   *Sol.*   *Res.*   *Sol.*   *Res.*

Pb	Bi Cd Cu NH$_4$OH.	As	Sb and Ag	Sn	Au Pt HCl and HNO$_3$.

Prec.   *Sol.*   *Sol.*

Bi	Cd Cu KCy and H$_2$S			Au	Pt

Prec.   *Sol.*

Cd.	Cu

(vertical labels, left to right):
Dissolve in aqua regia and test with SnCl$_2$.
White.   White.   Yellow.   HNO$_3$ + K$_4$Fe Cy, Red.   See page 202.   Test with HgCl$_2$.   SnCl$_2$, Purple.   KCl, Yellow.

COMPLETE ANALYSIS.

FILTRATE.

Add H_2S.

FILTRATE.

Add NH_4OH and $(NH_4)_2S$.

Prec. | *Filtrate.*

Co Ni Fe Cr Mn Al Zn
Dilute HCl.

Res. | *Sol.*

Co Ni.

Test with KNO_2, or borax bead, page 292.

Fe Cr Mn Al Zn
Boil in NaOH.

Prec. | *Sol.*

Fe Cr Mn
Fuse with KNO_3 and Na_2CO_3. Dissolve in H_2O.

Res. | *Sol.*

Fe — Dissolve in HCl. Add KCyS. Red.

Cr — Acetic Acid + Pb2NO_3. Yellow.

Mn — Green to red.

Al — Acidify with HCl and add NH_4OH, white.

Zn — H_2S. White.

Add $(NH_4)_2CO_3$.

Prec. | *Filtrate.*

Ba Sr Ca
Dissolve in Acetic Acid; 2 Parts.

I. | *II.*

I. — Add K_2CrO_4.
Prec. | *Sol.*

Ba — Yellow precipitate.

Sr — Add $CaSO_4$, white prec. in 30 min.

II. — Ca — H_2SO_4, filter; $NH_4OH + (NH_4)_2C_2O_4$, white.

Mg K Na
2 Parts.

I. | *II.*

I. — Mg — $NH_4OH + Na_2HPO_4$, white.

II. — Evaporate and test in flame; Na yellow, K violet.

Test for NH_4 in the original solution, by heating with NaOH, NH_3 is given off.

QUESTIONS FOR CLASS USE.

I.—INTRODUCTION.

Page 1, 2.—Define chemistry. What is the distinction between physical and chemical phenomena? Illustrate. Can matter be destroyed? What becomes of it when it disappears? What properties do gases possess which prove them to be matter? What is an element? How many are known? Is it probable that all the elements have been discovered? What is a compound? Are compounds like their elements? Illustrate. Define chemical affinity. Illustrate. How does it act?

3, 4.—What is the action of heat? Of light? Of electricity? Of solution? Illustrate. State the laws of definite and multiple proportions. What is the constitution of bodies? What is a molecule? An atom? How do atoms differ? What is atomic weight? Molecular weight? Valence?

5-7.—What notation is used in chemistry? Illustrate. What is the formula of a substance? Illustrate. How are reactions represented? Illustrate. How are elements named? Compounds? How are elements classified? What is the distinction between Inorganic and Organic Chemistry?

II.—INORGANIC CHEMISTRY.

1.—THE NON-METALS.

11. OXYGEN.—Give the symbol and atomic weight of oxygen. What is the meaning? Where is O found? How may it be prepared?

12.—How is O prepared from potassium chlorate and manganese dioxide? Give the reaction. What becomes of the potassium chloride which is formed?

13-16.—What is the use of the black oxide of manganese? Name the properties of O. What is oxidation? An oxide?

Show that O is a supporter of combustion. What compounds are formed in these illustrations? Describe the action of the O in the air. Describe the action of O on fuel. On impure water. On writing-ink. On red-hot iron. On damp knives and forks. By what means is the O carried through the system?

17, 18.—What work does O perform in the body? Why is the blood in the arteries red and in the veins blue? What chemical processes are included by the chemist under the term oxidation? Does fire differ essentially from decay? Is heat always produced by oxidation? Illustrate. What is the igniting point? How are fires extinguished? What causes spontaneous combustion?

19, 20.—What is the chemical process of starvation? Why does unusual exercise cause one to breathe more rapidly? Why does running cause panting? Why do we need extra clothing when we sleep, even at midday, in the summer? How do hibernating animals illustrate this? How does a cold-blooded animal differ from a warm-blooded one? How does O give us strength? What is potential energy? Kinetic energy?

21, 22.—Show how O is constantly burning the body. Is there any part of the body that is permanent? Illustrate the rapidity of this change. Show the truth of the paradox—"*We live only as we die.*" Why do we need food and sleep? Show how O acts as a scavenger in nature. In what sense is O the sweeper of the body? Is this a useful provision? How much O does each adult need per day? Total amount used daily?

23, 24.—What would be the result if the air were pure O? What objects would escape combustion? What is ozone? Where is it noticed? Preparation? Test? Properties? Is it a valuable constituent of the air? How does the ozone molecule differ from that of O?

25, 26.—What are the laws for the effects of change of temperature and pressure on gases? How can the weight of any volume of gas be calculated? How can the weight of a gas produced by a given weight of re-agents be found?

27, 28. NITROGEN.—Symbol and atomic weight? Why so called? Sources? Occurrence? Preparation? Properties? Why does a person drown in water? Would a person die in pure N? What is the peculiarity of the nitrogen compounds?

29, 30.—What causes flesh to decompose so much more easily than wood? Does the N we take in at each breath do us any direct good or harm? Where do we get N to make our flesh? Describe the action of N and O in our stoves. Where do plants obtain N? State the main distinction between O and N. What is the office of the N in the air? Show that the proportion of O and N in the atmosphere gives us the golden mean.

31, 32.—Formula and molecular weight of nitric acid? Explain its occasional presence in the atmosphere. Preparation? Properties? What color does it give to wood? Uses? Illustrate its oxidizing action. What is aqua regia? Describe the process of etching. The action of HNO_3 on Sn. What are the red fumes which pass off?

33, 34.—Formula and molecular weight of nitrous oxide? The common name? Preparation? Reaction? Properties? What is the effect of nitrous oxide on the human system? State its use in surgical operations. Formula and molecular weight of nitric oxide? Its preparation? Why is the gas in the flask colored? What compound is formed with O? Properties of NO? What are the fumes which it forms in the air?

35-37.—Formula and molecular weight of ammonia? Why so called? Its old name? What is aqua ammonia? Whence obtained? Give the reaction. Properties? How liquefied? Define the nascent state.

38, 39. HYDROGEN.—Symbol and atomic weight? Meaning of the name? Occurrence? Preparation? Reaction? What compound is formed? Properties? Is H a metal? *Ans.*—In all reactions it plays the part of a metal, and, like most of the metals, is electro-positive. Its levity? Will it destroy life? Effect on the voice? Use in filling balloons?

40-44.—What is the product of the combustion of H? What becomes of it? Will the gases H and O combine, if mixed? Describe the hydrogen gun. Cause of the report? What is the action of platinum sponge on a jet of H? Describe Dobereiner's lamp. Explain the heat produced by burning H.

How are hydrogen tones produced? Explain.

45-47. WATER.—Formula and molecular weight of water? How may its composition be proved? What is the freezing and the boiling point of water? What injury may a small quantity of H_2O do, if thrown on a fire? Explain. Can H_2O, then, be burned? Illustrate the abundance of H_2O in the animal world. Vegetable world. Mineral world. Why will blue vitriol lose its color if heated? What is "burnt alum"? Water of crystallization?

48-52.—Show the adaptation of H_2O as a solvent. What water is the purest? Why does rain-water taste so insipid? Is river-water a healthy drink? What is hard water? Soft water? Why does the hardness of water vary in different places? Is hard water healthful? How may we detect organic matter in H_2O? What minerals are most common in water? What is the "fur" in a tea-kettle? Why does soap curdle in hard water? How could Salt Lake be freshened? What is the use of the air in H_2O? How do fish breathe? Does the air in water differ from ordinary air? How? Why is boiled water so insipid? Give some of the paradoxes of water. Name the various uses of water. ("Physics," p. 201.)

53-55. CARBON.—Symbol and atomic weight? Illustrate the abundance of C. Is it more characteristic of the vegetable than of the mineral kingdom? What are its forms? Proof of these allotropic states? What is an allotropic condition? What is the diamond? Properties? Has it ever been made artificially? What is a carat? How is the diamond ground? Describe the three modes of cutting. What gives the diamond its value? Common name for graphite? Origin? Uses of graphite?

56, 57.—Describe the process of making a lead-pencil. What is a black-lead crucible? What is British Luster? Lamp-black? Uses? Fitness for printing? What can you say about ancient MSS.? What is soot? What causes the burning of chimneys? Does this occur oftener when wood than when coal is used as a fuel? How is charcoal made? What is coke? Uses? Gas-carbon?

58, 62.—Bone-black? Uses? How is sugar refined? Describe the formation of coal. Difference between bituminous and anthracite coal? Why is coal found in layers, with slate, etc., between? What proof have we that coal is of vegetable origin? Describe the formation of peat. Uses? What is muck? Use? Name some of the diverse properties and uses of C.

63. CARBON DIOXIDE. — Formula and molecular weight? Occurrence? How is it constantly formed? Preparation? Reaction?

64, 65.—Test? What causes the pellicle on lime-water? What does this show? Prove that we exhale CO_2. Give the properties of CO_2. Prove that CO_2 is heavier than air. A non-supporter of combustion. That it contains C.

66, 67.—What test should be employed before descending into a deep well or an old cellar? How can you remove the foul air? Tell about the Grotto del Cane. Is CO_2 directly poisonous? What is choke-damp? Fire-damp? Which is more dreaded?

68, 69.—Has CO_2 been used in extinguishing fires? Tell about the absorption of CO_2 by H_2O. What is soda-water? How is CO_2 liquefied? Why does the liquid solidify when exposed to the air? What principle in natural philosophy does this illustrate? How low a degree of cold has been produced in this manner? Describe the need of ventilation. How is the air expired from our lungs made useful? Is a single opening sufficient to ventilate a room? What practical application do you make of this subject?

70, 71.—Formula and molecular weight of carbon monoxide? Properties? Where is it often formed? Explain. Practical importance of this fact? What causes the unpleasant odor of coal-gas? Formula and molecular weight of light carburetted hydrogen? Properties? How is it formed?

72, 73.—Name places where it is found in great quantities. Formula and molecular weight of heavy carburetted hydrogen? Properties? How made? What gases mainly compose coal-gas? Which is the most valuable? Describe the manufacture. Is the odor an advantage? Is coal-gas explosive? Why is the jet flat? When we turn the gas very low, or the supply is insufficient, why is the flame blue?

74, 75.—What is water-gas? Formula and molecular weight of cyanogen? Meaning of the name? Preparation? What are its compounds called? What is the yellow prussiate of potash? The red? *Ans.*—The ferricyanide, K_3FeCy_6. Properties of Cy? What is a compound radical? Formula of hydrocyanic acid? Common name? Where found? Antidote? What are the fulminates? How are gun-caps made?

76, 77. COMBUSTION.—Define. What is a combustible? A supporter of combustion? (The difference between these two is nicely shown in the experiment with H on p. 40.) A burnt body? *Ans.*—A body which has combined with O.—*Example:* a stone, water. Upon what does the amount of heat produced by combustion depend? The intensity? Why do we need a draught to a stove? Does combustion, in its chemical sense, commence before the fuel catches fire? Why do we use "kindlings" in starting a fire? Why can we light pitch-pine so easily? What are hydrocarbons? What are the ordinary products of combustion? What causes the dripping of stove-pipes? What are the ashes? Why does fresh fuel produce a flame? Show how admirably C is adapted for a fuel. What would be the effect if CO_2 were not a gas? Define flame. Describe the burning of a candle. Show that flame is hollow.

78, 79.—What causes the light? Why is the flame blue at the bottom? Products of combustion? Tests? Why does the wick turn black? What causes the coal at the end of the wick? Why does snuffing brighten the light? Why does a draft of air, or a sudden movement of the candle, cause it to smoke? Why is the flame of a candle or lamp red, or yellow? *Ans.*—Because the heat is not sufficient to cause the carbon to emit all the rays of the spectrum. Use of plaited wicks? Object of a chimney to a lamp? A flat wick? Advantage of an Argand lamp? What is the film which gathers on the chimney when the lamp is first lighted? Why does this soon disappear? Why do tar, spirits of turpentine, etc., burn with much smoke?

80-82.—Why does alcohol give much heat and no smoke? Describe Davy's safety-lamp. Illustrate this by a wire gauze over the flame of a candle. Describe Bunsen's burner. Why does it give great heat, little light, and no smoke? Describe the oxy-hydrogen blow-pipe. Why does it give great heat and little light? What is the calcium light?

83.—Describe the mouth blow-pipe. The blow-pipe flame. Where is the hottest point in the flame? What is the reducing flame? The oxidizing flame? Why does blowing on a candle-flame extinguish it?

85-89. THE ATMOSPHERE.—Name the constituents. Proportion. State the comparison. What is diffusion? What effect

does this have on the air? Is the air a chemical compound?
Illustrate. Has each constituent a special use? Name the uses
of O. Of CO_2. Explain the chemical change which takes place
in the leaf. What is the influence of house-plants upon the
atmosphere of a room? What can you say of the exact balance
kept between the wants of animals and plants?

90, 91.—What relation exists between animals and plants?
Which gathers and which spends the solar energy? Which performs the office of reduction? Which that of oxidation? How
is the solar energy set free? What is the use of the watery
vapor in the air? Which of the constituents are permanent?
Is this a wise provision? What effect does this permanence
have upon sound? Why ought the vapor to be easily changed
to the liquid form?

92. THE HALOGENS.—Name them. Symbols and atomic
weights? Compare the halogens with each other. What compounds do they form? Why is chlorine so called? Occurrence?
Preparation? Reaction?

93-95.—Properties of Cl? What action does Cl have on phosphorus, arsenic, etc.? Why does a solution of the gas soon
become acid? What is its action on organic bodies? Why does
Cl act more readily in the presence of moisture? Its action on
turpentine? On printers' ink? Describe the chemical change
in domestic bleaching. The method of bleaching on a large scale.
What is the advantage of using Cl over other disinfectants?

96, 97.—How may the gas be set free? How are hospitals
purified? What mixture would liberate Cl in the greatest quantities? Formula and molecular weight of hydrochloric acid?
Common name? Preparation? Reaction? Properties? What
are its compounds termed? Tests? What is aqua regia?

98-100.—What is an acid? A base? A salt? How are acids
and salts named? Illustrate. Tell what you can about bromine. Its uses.

101, 102.—Why is iodine so called? Its source? Properties?
Test? What is the peculiarity of fluorine? Occurrence? What
acid does it form? For what is this acid noted? Describe the
process of etching with HF. Why is not HF kept in ordinary
bottles? Is it dangerous to use?

103. SULPHUR.—Symbol and atomic weight? Sources? What

is the principle of hair-dyes? Why do eggs tarnish silver spoons? What is the difference between brimstone and flowers of sulphur? Properties? Solvent? Allotropic forms? Describe the changes produced by heating.

104, 105.—Uses of S? Formula and molecular weight of sulphur dioxide? Where is it familiar? What is sulphurous acid? What salts does it form? Uses of SO_2 in bleaching? Why are new flannels liable to turn yellow when washed? Formula and molecular weight of sulphuric anhydride? By what other name is it known? Preparation? Properties? Why is Nordhausen acid so called?

106, 107.—Formula and molecular weight of sulphuric acid? Common name? State its importance. What are its compounds called? Illustrate the making of H_2SO_4. Describe its manufacture. Reaction?

108-110.—Properties? What especial property? Illustrate. Its strength? Color of its stain on cloth? How removed? On wood? Cause of this action? Test? Formula and molecular weight of hydrogen sulphide? Where is it found? Preparation? Reaction? Properties? Use? Color of the precipitates? Test? Formula and molecular weight of carbon sulphide? Preparation? Properties? Uses? How does it illustrate the force of chemical affinity?

111, 112. VALENCE.—What is valence? Illustrate. What names distinguish the different valence of atoms? What are monobasic, bibasic, tribasic acids? What is a normal salt? An acid salt?

113. PHOSPHORUS.—Symbol and atomic weight? Why so called? Occurrence? In what parts of the body, and in what forms, is it found?

114-116.—Preparation? Properties? Caution to be observed? Is phosphorus poisonous? What is the product of its combustion? Describe the amorphous form of phosphorus. What is the principal use of phosphorus? Describe the making of the lucifer match. The safety match. What compounds are formed in the burning of a match? What is phosphorescence? Its cause?

117.—Formula and molecular weight of hydrogen phosphide? Preparation? Properties?

312 QUESTIONS FOR CLASS USE.

117-120. ARSENIC.—Symbol and atomic weight? Common name? Test? What is commonly sold as arsenic or ratsbane? Preparation of arsenic trioxide? Properties? What can you tell of its antiseptic property? Antidotes? Describe Marsh's test. How can the As be distinguished from Sb? What is said of arsenic eating?

120-122. BORON.—Symbol and atomic weight? Source? Describe the scene in Tuscany where it is found. Process of obtaining boric acid? Formula and molecular weight of borax? Uses?

123-126. SILICON.—Symbol and atomic weight? Occurrence? Common names of SiO_2? What gems does it form? What is sand? Properties? Why is it called an anhydride? Is silica soluble in H_2O? How does it get into plants? In what plants is it found? Explain the process of petrifaction. What is said of the antiquity of glass? Pliny's story of its origin? What is said of its value in the twelfth century? Name the four varieties of glass and the composition of each. What are the essential ingredients of glass? How is glass colored? Name the oxides used. Why is flint-glass so called? How is glass annealed? Describe the Prince Rupert's drop. How are Venetian balls made? Tubes? Beads?

2.—THE METALS.

127. POTASSIUM.—Symbol and atomic weight? History of its discovery? Source?

128, 129.—How do we get our supply? Preparation? Properties? How must it be kept? Reaction when thrown on H_2O? Formula and molecular weight of caustic potash? Properties? Its feel? Its affinity for H_2O? Uses? Formula and molecular weight of potassium carbonate? Common name? Preparation? What part of the tree furnishes the most potash? What is the derivation of the word? Formula and molecular weight of acid potassium carbonate? Common names? Preparation? Formula and molecular weight of potassium nitrate? Common names? Where is it found?

130, 131.—How is it prepared artificially? How much water would be required to dissolve a pound of this salt? Properties?

Uses? What is the composition of gunpowder? Cause of its explosive force? Uses of potassium chlorate? Potassium bichromate? Composition of fire-works?

SODIUM.—Symbol and atomic weight? Source? What proportion does it form of common salt?

132, 133.—What element does it resemble? Reaction when thrown on water? What compound is formed? Test? Formula and molecular weight of common salt? What use does it subserve in the body? Is salt abundant? Describe the manufacture. What is solar salt? Describe the "hopper-shape" crystal. Is it best to heat the water for dissolving salt? What is a saturated solution?

134, 135.—Formula and molecular weight of sodium hydroxide? Common name? Uses? Formula and molecular weight of sodium sulphate? Common name? Preparation? Reaction? What curious property has this salt? Why will the dropping in of a crystal cause solidification? Formula and molecular weight of sodium carbonate? Common names? Why called carbonate of soda? Describe its manufacture. Why will Na_2CO_3 soften hard water? Formula and molecular weight of acid sodium carbonate? Common name? Why called bicarbonate of soda? Preparation? Use?

136, 137.—Give the theory of ammonium. How is the formula NH_4OH obtained? What is a compound radical? Formula and molecular weight of ammonium chloride? Preparation? Uses? Formula and molecular weight of ammonium carbonate? Common names? Uses? Formula and molecular weight of ammonium nitrate? Preparation? Uses?

138, 139. CALCIUM.—Symbol and atomic weight? Source? In what part of the body is it found? In what form do we commonly see it? Formula and molecular weight of lime? Preparation? Describe a lime-kiln. Properties of CaO? Test? What is the difference between "water-slaked" and "air-slaked" lime? Uses? What is whitewash? Concrete? Hard-finish? Calcimining? Theory of the hardening of mortar? Why are newly-plastered walls so damp? Will mortar harden if protected from the air?

140, 141.—Action of lime on the soil? Will it not lose its beneficial effect after a time? Should it be applied to a compost heap? How can this waste be avoided? How would you test for the escaping NH$_3$? Action of lime on copperas? How does the copperas get in the soil? Uses of lime? Formula and molecular weight of carbonate of lime? Occurrence? How are stalactites and stalagmites formed? What is petrified moss? Whiting? Marble? Chalk? Marl? Formula and molecular weight of calcium sulphate? Common names? What is plaster of Paris? Why does plaster of Paris harden, if moistened? *Ans.*—Because it absorbs water again.

142, 143.—Uses? What is plaster? How prepared for use as a fertilizer? *Ans.*—It is ground into a fine powder. Tell the story of Franklin. What is the difference between sulphate and sulphite of lime? Formula and molecular weight of phosphate of lime? What is the superphosphate? Use? Uses of the salts of barium and strontium? What is heavy spar? Barytes?

144. MAGNESIUM.—Symbol and atomic weight? Occurrence? How can you tell if a stone contains Mg? *Ans.*—It generally has a soapy feel. Properties? For what is it noted? Product of its combustion? Formula and molecular weight of magnesium sulphate? Common name? What is magnesia alba?

145. ALUMINIUM.—Symbol and atomic weight? Occurrence? Properties? Solvents? What can you say of its abundance and probable usefulness? What is alumina? What crystals and gems does it form?

146, 147.—What is emery? What is common clay? Use in the soil? In the arts? What is ochre? Fuller's earth? Explain the process of glazing pottery ware. What is the salt glaze? The litharge glaze? What objection to the latter? What gives color to brick? What is the peculiarity of white brick? How is alum made? Name different kinds of alum. Which kind is the common commercial alum? Use of alum in dyeing? How are alum crystals made? *Ans.*—They are obtained by suspending threads in a saturated solution of this salt. In this manner alum baskets, bouquets, etc., are formed of any desired color.

148.—What is spectrum analysis? Is it a reliable test? Illustrate its delicacy. What is the spectroscope?

150. IRON.—Symbol and atomic weight? Tell what you can of its value to the world. How is its use a symbol of a nation's progress?

151, 152.—State how its value is enhanced by labor. Name the sources of iron. Common ores. Describe the process of smelting iron ore. Why is hot air used for the blast? Reaction of the lime? What becomes of the O in the ore?

153.—Origin of the term "pig-iron"? Name the varieties of iron. Difference between them. What is cast-iron? Its properties? Uses? How is iron adapted for castings? What is chilled iron? Wrought iron?

154.—Preparation? Effect of jarring? Illustrate its malleability. What is steel? Preparation? In making steel tools, how does the workman judge of the temper? How are cheap knives made?

155.—Describe Bessemer's process. Cause of the changing colors often seen in the scum over standing water?

156, 157.—Name the different oxides of iron. Give the formula of each. What peculiar property is possessed by the ferric oxide and ferric hydroxide? What is iron carbonate? By what name is it known? Cause of the ferruginous deposit around chalybeate springs? Formula and molecular weight of iron disulphide? Common names? What is chameleon mineral?

158, 159.—Uses of iron disulphide? Formula and molecular weight of ferrous sulphate? Common names? Uses?

ZINC.—Symbol and atomic weight? Source? Preparation? Reaction? Is it malleable? Will it oxidize in the air? Uses? What is philosopher's wool? What is galvanized iron? Are water-pipes made of this material safe? Formula and molecular weight of zinc oxide? Use? Formula and molecular weight of zinc sulphate? Use?

160. TIN.—Symbol and atomic weight? Where found? Properties? What is the "tin cry"? What is common tinware? Action of HNO_3 on Sn? What can you say of the manufacture of pins?

161. COPPER.—Symbol and atomic weight? Where found? Antiquity of the mines? What is malachite? Properties of Cu? Color of its vapor? Solvent? Test? What is verdigris?

162, 163.—Black oxide of copper? What is the danger of using a copper kettle? Formula and molecular weight of copper sulphate? Common name? Uses?

LEAD.—Symbol and atomic weight? Source? Preparation?

316

QUESTIONS FOR CLASS USE.

Properties? Its effect on the human system? Action of water
on lead? Is there more danger with hard, or with soft water?
What precaution should always be used with lead pipes? What
is the test of lead? What is "litharge"?

164.—Its uses? "Red-lead"? Its uses? What is "white-
lead"? Describe its manufacture. With what is it adulterated?
What is "sugar of lead"? Properties? Antidote? Explain the
formation of the lead-tree.

165, 166. GOLD.—Symbol and atomic weight? Source?
Preparation? What is an amalgam? Quartation? Properties?
Solvent? Process of making gold-leaf?

167-172. SILVER.—Symbol and atomic weight? Source?
Preparation, 1, from the sulphide; 2, horn-silver; 3, lead?
Describe the process of reduction at the West. What is cupellation? Properties of silver? Solvent? Test? What is the
common name of nitrate of silver? What is its action on the
flesh? How may its stain be removed? Uses? Of what are
hair-dyes and indelible inks made? Describe the process of
Daguerreotyping. Photography.

173. PLATINUM.—Symbol and atomic weight? Source? Preparation? Properties? Uses? How is very fine platinum wire
made?

174-176. MERCURY.—Symbol and atomic weight? Common
name? Why so called? Source? Preparation? Properties?
Uses? Action on the human system? Process of silvering
mirrors? What is blue-pill? Mercurial ointment? Formula of
mercuric oxide? Mercurous chloride? Mercuric chloride?
Mercuric sulphide? Common names? Uses? Properties?

THE ALLOYS.—What is an alloy? What peculiarity with regard to the melting point? Of what is type-metal made?

177.—Pewter? Britannia? Brass? German silver? Solder?
Fusible metal? Bronze? How is gold soldered? Silver? Copper?

178.—What is the principle? What are the constituents of
gold coin? Silver coin? What is the meaning of the term
carat? How are shot manufactured? How are they sorted?

179-180.—What is or-molu? Aluminium bronze? Compare
the properties of the metals with regard to, 1, oxidation; 2,
density; 3, melting point; 4, color; 5, malleability; 6, brittleness; 7, tenacity; 8, special properties.

III.—ORGANIC CHEMISTRY.

185-189. INTRODUCTION.—What is organic chemistry? What was the first organic substance artificially made? Name some which have since been made. What are the organized bodies? What elements do organic substances contain? Explain the great number of organic substances. What is isomerism? Are organic molecules often complex?

189-192. THE PARAFFINES.—What is the general formula of this series of hydrocarbons? Name some of the members. Why is the series so called? Source of petroleum? How is it purified? What are the products of its distillation? Danger of kerosene explosions. How can the kerosene be tested? What is paraffine? Bitumen? Its properties? Uses? How may the paraffines be artificially made?

193, 194. THE ALCOHOLS.—What is an alcohol? Formula of methyl alcohol? Its source? Its uses? Ethyl alcohol—formula? Source? Uses? Effects on the human system?

195-199. FERMENTATION.—Cause? Does it ever take place spontaneously? How does the yeast act? What change takes place in the alcoholic fermentation? The acetic? Describe the formation of yeast. The making of malt. Yeast cakes. What is gluten? How does it act? What is diastase? Describe the brewing of beer. Why is lager beer so called? Describe the making of wine. What is the difference between a dry, a sweet, and an effervescing wine? Cause of the flavor? State the proportion of alcohol in common liquors. How is brandy made? Rum? Whisky? Gin? Describe the apparatus used for distillation. What is fusel oil?

200-204. THE ALDEHYDES AND ACIDS.—What is an aldehyde? Formula of ethyl aldehyde? How is it formed? Formula of formic acid? Occurrence? From what is it made? Formula of acetic acid? What is glacial acetic acid? How is vinegar made? What is cider vinegar? Pyroligneous acid? Properties of acetic acid? Use? What causes the working of preserves? Where is oxalic acid found? Preparation? Properties? Antidote? Uses? Where is malic acid found? Citric? Where is tartaric acid found? Preparation?

What is cream of tartar? Tartar emetic? Rochelle salt? Seidlitz powders?

204-209. THE ETHERS AND ETHEREAL SALTS.—What is an ether? Formula of ordinary ether? Why called "sulphuric" ether? Its properties? Uses? What are ethereal salts? Illustrate. What are the principal fats? Formula of glycerin? To what class of organic substances does it belong? From what source and how is it obtained? Properties? What is nitro-glycerin? Dynamite? How are candles made? Illustrate the formation of soap. What is the reaction? Difference between hard and soft soap? What is the cause of the curdling of soap in hard water? Describe the cleansing action of soap. What is saponification?

209, 210. THE HALOGEN DERIVATIVES.—Formula of chloroform? How made? Properties? Iodoform? Chloral?

211-213. STARCH.—Formula? Sources? Use in the plant? Why stored in that form? Appearance under the microscope? Preparation? Properties? What is dextrin? Test of starch? Varieties? What is gum? Composition? Mucilage? Is it soluble in water? What is pectose? Pectin?

213-217. CELLULOSE.—Formula? What is the composition of wood? Name the various forms of cellulin. Illustrate the wonders of secretion. State the uses of woody fiber. The making of paper. Paper-parchment. Linen. Cotton. Guncotton. Collodion. Its uses.

217-220. SUGAR.—Cane-sugar? How is sugar refined? Difference between loaf and granulated sugar? Describe a centrifugal machine. What is terra alba? Use? Of what are gumdrops made? Rock-candy? What is caramel? Use? Formula of grape-sugar? Source? Sweetening power? How is sugar made from starch? How does the oil of vitriol act? How do jellies, preserves, etc., "candy"? Why are dextrose and levulose so named? Why must matter be organized? What is the office of plants?

221-225. THE AROMATIC COMPOUNDS.—Formula of benzene? Its source? Properties? How is nitro-benzene made? Its formula? Uses? Formula of aniline? How made? Properties? What is carbolic acid? Picric acid? Naphthalene? Anthracene? Benzoic acid? Salicylic acid? Benzoic aldehyde? Toluene?

225-231. THE TERPENES AND CAMPHORS.—How do the volatile oils differ from the fixed oils? Sources of the essential oils? Their preparation? Their composition? Formula of oil of turpentine? Its properties? What is camphene? Burning fluid? Camphor? Its preparation? ˙Properties? What are the resins? The balsams? Illustrate. What is the source of resin? Its uses? What is lac? Difference between stick-lac, seed-lac, and shellac? How is sealing-wax made? Source of gum-benzoin? Uses? Amber? Origin? Properties? Uses? India-rubber? Source? Properties? Uses? What is vulcanized rubber? Properties? Gutta-percha? Uses?

231-234. THE ALKALOIDS.—Sources? What is opium?˙ Preparation? Uses? Laudanum? Paregoric? Danger of opium-eating? What is morphine? Use? Quinine? Use? Nicotine? Properties? Strychnine? Properties? The chromatic test? Name the active principle of tea and coffee. What substances are found in tea? In coffee? Describe the process of tea-raising. Of making black tea. Green tea.

235-239. DYES AND DYEING.—Source of organic coloring principles? What is an adjective color? A substantive color? A mordant? The process of dyeing? Of calico printing? What is madder? Its coloring principle? Cochineal? Use? Brazil-wood? Use? Indigo? Preparation? White indigo? Logwood? Litmus? Leaf-green? Tannin? Name its varieties. What are nut-galls? Properties of tannin? Describe the process of tanning. How is leather blackened? How is ink made? Why does writing-fluid darken by exposure to the air? What is gallic acid? Pyrogallic acid? Use?

239-243. ALBUMINOUS BODIES.*—What is their composition?

* Notice here the wise provision of nature. Nitrogen, slow and sluggish when uncombined, is fitted to dilute the air; while N, restless and uneasy when combined, is equally adapted to form unstable compounds of food, to carry force into our bodies and there to quickly set it free. Oxygen, when free, is active, eager, and ready to search the nooks and crannies of the capillaries; but when once it combines with a substance, takes it for better or for worse, and forms the stablest of compounds. We find nitrogen compounds in the animal and vegetable worlds, ready for use where they are needed, in our muscles. Oxygen compounds are abundant in the mineral world, and stored in the seeds of plants, at hand to give form to the more permanent parts of the body. Such profound relations, such nice adapta-

What is albumin? Source? Properties? Casein? Why does milk curdle? Action of rennet? Why does cream rise on milk? Describe the souring of milk. Fibrin? Properties? Gluten? Legumin? Putrefaction? Cause? Why does salt preserve meat? What is gelatin? Glue? Isinglass? Size?

243-247. DOMESTIC CHEMISTRY. — Describe the chemical changes which take place in making bread. What is stale bread? Why is it dry? How is aerated bread made? Why is bread ever sour? How are griddle-cakes raised? Biscuit? What are baking-powders? Action of soda and HCl? Of sal-volatile? How is bread changed by toasting? How are potatoes changed by cooking?

tions of our bodies to the world around, give us glimpses of a creative skill worthy of our noblest thought and highest admiration.

INDEX.

This Index includes the Notes as well as the Text.

GLOSSARY.

A

ăç′e tātᶒ
a çĕt′ ie
ae tĭn′ ie
ăf fĭn′ i ty
al bū′ min
al bū′ mi noŭs
ăl′ eᶅe mist
ăl′ de hȳdᶒ
a lĭz′ a rĭn
ăl′ ka lĭ
ăl lo trŏp′ ie
al loy′
al ū mĭn′ i um
a măl′ gam
am mŏ′ ni à
am mo nĭ′ ae al
a môr′ phoŭs
ăm′ ȳl
ăn æs thĕt′ ie
ăn hȳ′ drĭdᶒ
ăn hȳ′ droŭs
ăn′ ĭ lĭnᶒ
ăn′ thră çēnᶒ
ăn′ tĭ mo ny
ä′ quà
är ǵĕn′ tum
är′ se nie
är se nĭ′ ŭ rĕt ted
är′ ter y
ăs phȳx′ i à

ăt′ om
a tŏm′ ie
ạu′ ǵītᶒ

B

bā′ rĭ um
ba rȳ′ tēs
bĕn′ zēnᶒ
ben zō′ ie
bĕn′ zŏl
bĕr′ ȳl
bĭ bā′ sie
bĭ eär′ bo nātᶒ
bĭ′ na ry
bĭ nŏx′ ĭdᶒ
bĭ tū′ men
bĭ tū′ mi noŭs
bĭv′ a lent
bŏ′ rax
bŏ′ rie
bŏ′ ron
brŏ′ mĭnᶒ
brụ′ çĭnᶒ
bū′ tȳ rātᶒ
bu tȳr′ ie

C

eæ′ ṣi ŭm
eăf fē′ ĭnᶒ
eăf fe o tăn′ nie
eăl eā′ re ŏŭs
eal çĭnᶒd′

eăl′ çĭ um
eăl′ o mĕl
eăm′ phēnᶒ
eaạut′ chọue
(kōo′ chōok)
eăp′ il la ry
eär′ bŏ hȳ drātᶒ
eär bŏl′ ie
eär bo nā′ ceoŭs
eär′ bon ātᶒ
eär bŏx′ ȳl
eär′ bu rĕt
eär′ bu rĕt ted
eär′ mĭnᶒ
eā′ se ĭn
eạus′ tie
çĕl′ lu lĭn
çĕl′ lu loid
çĕl′ lu lōsᶒ
ehal çĕd′ o ny
eha lȳb′ e ātᶒ
eᶅĕm′ iṣm
eᶅĕm′ is try
eᶅlō′ ral
eᶅlō′ rĭnᶒ
eᶅlō′ ro form
eᶅlō′ ro phȳl
eᶅrŏ măt′ ie
eᶅrȳs′ o prạṣᶒ
çĭn′ na bar
çĭt′ rie

eŏ'balt
eŏch'i něal
eol lŏ'di on
eŏp'per as
erē'o sōtę
eu pěl'
eŭ pel lā'tion
çȳ'a nĭdę
çȳ ăn'o ġen
çȳm'o ġēnę

D

de ŏx'i dĭzę
děx'trĭn
děx'trŏşę
dĭ'as tāsę
dĭ ŏx'ĭdę
dĭ sŭl'phĭdę
dŏl'o mĭtę
dȳ'na mĭtę

E

ěb'on ĭtę
ē lee trŏl'y sĭs
ěth'yl

F

fěr'rie
fĭ'brĭn
fĭ'eus
flŭ'ŏr ĭnę
fŭl'mĭ nātę
ful mĭn'ie
fŭ'sel

G

ga lě'nä
găl lo-tăn'nie
găs'o lĭnę
gaş ŏm'e ter
ġěl'a tĭn

glŏb'ūlę
glŭ'ten
glȳç'er ĭn
grăph'Itę

H

hăl'o ġen
hä'loid
hěm'a tĭtę
hôrn'blěndę
hȳ drau'lie
hȳ dro brŏ'mic
hȳ dro eär'bon
hȳ dro eḩlŏ'rie
hȳ dro çȳ ăn'ie
hȳ'dro ġen
hȳ drŏx'ȳl
hȳ dro zō'ȧ
hȳ po eḩlōr'Itę

I

I'o dĭdę
I'o dĭnę
I ŏd'o fôrm
I so měr'ie
I sŏm'er Işm
I so môrph'oŭs

J

jär go něllę'

K

kělp
kěr'o sēnę
kĭ nět'ie

L

lăe
lau'da nŭm
le gū'mĭn

lĭ'ehen
lĭth'arġę
lĭth'i ŭm
lĭt'mus

M

măg ně'si ŭm
măl'a eḩĭtę
mä'lie
măŋ ga něşę'
mauve (mōv)
meer'sḩaŭm
měr eū'rie
měr' eū roŭs
měr'eu ry
mĭr'bānę
mo lěe'ū lar
mŏl'e eūlę
mŏn o bā'sie
mo nŏx'Idę
môr'dant
môr'phĭnę
mū ri ăt'ie

N

năph'thä
nĭe'o tĭnę
nĭ'trātę
nĭ'ter
nĭ'trie
nĭ'tro ġen

O

ŏ'eḩry
ŏe ta hě'dral
œ năn'thie
ŏ'le fĭ ant
ŏ'le ĭn
ŏ'pi ŭm

ôr mo lụ́
ox ăl′iе
ŏx i dā′tion
ŏx′y ģĕn
ŏx y hȳ′dro ģen
ō′zōnᶒ

P

pal mĭt′iе
păl′mi tĭn
păr′af fĭnᶒ
.pĕе′tōsᶒ
pen tŏx′Idᶒ
per măṇ′gan ătᶒ
phē′nol
phē′nyl
phŏs′phĭnᶒ
phŏs′phu rĕt ted
pĭ′еriе
plăt′i nŭm
plum bā′go
po tăs′si ŭm
prō′te In
prŭs′si ātᶒ
prŭs′siе
pȳ rĭ′tᶒş
pȳr o găl′liе
pȳr o lĭg′ne oŭs
pȳ rŏx′y lĭn

Q

quạd rĭv′a lent
quĕr çi tăn′niе
quĭl′nĭnᶒ
quạrtz

R

rĕş′in
rh̩Ig′o lēnᶒ
rŏş ăn′i lĭnᶒ
rŏş′in
ru bĭd′i ŭm

S

săl-am mō′ni ae
săl e rā′tus
săl I çȳl′iе
sa pŏn i fi еā′tion
săr′do nȳx
sᶒĭd′litz
sĕl′e nĭtᶒ
sĕs quI еăr′bo nātᶒ
sĕs quI ŏx′Idᶒ
shĕl′lăе
sĭ′lex
sĭl′i еātᶒ
sĭl′i еon
smạlt
sō′di ŭm
sôr′gh̩um
spăth′iе
spĕе′ū lŭm
spĕrm a çē′tĭ
spŏr′ūlᶒ
sta lăе′tĭtᶒ
sta lăģ′mĭtᶒ
stăn′niе
stăn′noŭs
ste ăr′iе
stē′a rĭn
stŏm′a tá

strŏn′ti an Itᶒ
strŏn′tĭ ŭm
strȳеh̩′nĭnᶒ
sū′еrōsᶒ
sŭl′phur
sŭl′phu rĕt ted
sulph ȳ′driе

T

taṛ tăr′iе
tĕr′·pĕnᶒş
tĕt′a nŭs
thē′Inᶒ
thē i tăn′niе
tŏl′ụ ēnᶒ
tō lụ′I dïnᶒ
trI bā′siе
trI ō′le ātᶒ
trI păl mI′tătᶒ
trI stē′a rātᶒ
trĬv′a lent
tûr′pen tĭnᶒ

U

ŭ nĭv′a lent
ū′pas
ū′re á

V

văl′ençᶒ
vĕr′dĬ grĭs
vĭt′rĬ ol
vŭl′еan Itᶒ

Z

zăf′fer